Lecture Notes in Mathematics

Volume 2352

Editors-in-Chief
Jean-Michel Morel, City University of Hong Kong, Kowloon Tong, China
Bernard Teissier, IMJ-PRG, Paris, France

Series Editors
Karin Baur, University of Leeds, Leeds, UK
Michel Brion, UGA, Grenoble, France
Rupert Frank, LMU, Munich, Germany
Annette Huber, Albert Ludwig University, Freiburg, Germany
Davar Khoshnevisan, The University of Utah, Salt Lake City, UT, USA
Ioannis Kontoyiannis, University of Cambridge, Cambridge, UK
Angela Kunoth, University of Cologne, Cologne, Germany
Ariane Mézard, IMJ-PRG, Paris, France
Mark Podolskij, University of Luxembourg, Esch-sur-Alzette, Luxembourg
Mark Policott, Mathematics Institute, University of Warwick, Coventry, UK
László Székelyhidi, MPI for Mathematics in the Sciences, Leipzig, Germany
Gabriele Vezzosi, UniFI, Florence, Italy
Anna Wienhard, MPI for Mathematics in the Sciences, Leipzig, Germany

This series reports on new developments in all areas of mathematics and their applications - quickly, informally and at a high level. Mathematical texts analysing new developments in modelling and numerical simulation are welcome. The type of material considered for publication includes:

1. Research monographs
2. Lectures on a new field or presentations of a new angle in a classical field
3. Summer schools and intensive courses on topics of current research.

Texts which are out of print but still in demand may also be considered if they fall within these categories. The timeliness of a manuscript is sometimes more important than its form, which may be preliminary or tentative. Please visit the LNM Editorial Policy (https://drive.google.com/file/d/1MOg4TbwOSokRnFJ3ZR3ciEeKs9hOnNX_/view?usp=sharing)

Titles from this series are indexed by Scopus, Web of Science, Mathematical Reviews, and zbMATH.

Daniela Bubboloni • Pablo Spiga •
Thomas Stefan Weigel

Normal 2-Coverings of the Finite Simple Groups and their Generalizations

Daniela Bubboloni
Dipartimento di Matematica e Informatica
"U.Dini"
Università degli Studi di Firenze
Firenze, Italy

Pablo Spiga
Dipartimento di Matematica e Applicazioni
Università di Milano-Bicocca
Milano, Italy

Thomas Stefan Weigel
Dipartimento di Matematica e Applicazioni
Università di Milano-Bicocca
Milano, Italy

ISSN 0075-8434   ISSN 1617-9692 (electronic)
Lecture Notes in Mathematics
ISBN 978-3-031-62347-9   ISBN 978-3-031-62348-6 (eBook)
https://doi.org/10.1007/978-3-031-62348-6

Mathematics Subject Classification: 05E16, 20Dxx, 20B05, 05E18

© The Editor(s) (if applicable) and The Author(s), under exclusive license to Springer Nature Switzerland AG 2024

This work is subject to copyright. All rights are solely and exclusively licensed by the Publisher, whether the whole or part of the material is concerned, specifically the rights of translation, reprinting, reuse of illustrations, recitation, broadcasting, reproduction on microfilms or in any other physical way, and transmission or information storage and retrieval, electronic adaptation, computer software, or by similar or dissimilar methodology now known or hereafter developed.

The use of general descriptive names, registered names, trademarks, service marks, etc. in this publication does not imply, even in the absence of a specific statement, that such names are exempt from the relevant protective laws and regulations and therefore free for general use.

The publisher, the authors and the editors are safe to assume that the advice and information in this book are believed to be true and accurate at the date of publication. Neither the publisher nor the authors or the editors give a warranty, expressed or implied, with respect to the material contained herein or for any errors or omissions that may have been made. The publisher remains neutral with regard to jurisdictional claims in published maps and institutional affiliations.

This Springer imprint is published by the registered company Springer Nature Switzerland AG
The registered company address is: Gewerbestrasse 11, 6330 Cham, Switzerland

If disposing of this product, please recycle the paper.

Preface

Given a finite group G, we say that G has weak normal covering number $\gamma_w(G)$ if $\gamma_w(G)$ is the smallest integer with G admitting proper subgroups $H_1, \ldots, H_{\gamma_w(G)}$ such that each element of G has a conjugate in H_i, for some $i \in \{1, \ldots, \gamma_w(G)\}$, via an element in the automorphism group of G.

We prove that the weak normal covering number of every non-abelian simple group is at least 2, and we classify the non-abelian simple groups attaining 2. As an application, we classify the non-abelian simple groups having normal covering number 2. We also show that the weak normal covering number of an almost simple group is at least two, except for a very specific family of almost simple groups arising from 8-dimensional orthogonal groups having Witt defect zero.

We determine the weak normal covering number and the normal covering number of the almost simple groups having socle a sporadic simple group. Using similar methods, we find the clique number of the invariably generating graph of the almost simple groups having socle a sporadic simple group.

In the introductory section of this monograph, we give various motivations for this investigation; broadly speaking, besides the intrinsic theoretical interest in (weak) normal 2-coverings, in recent years, a few remarkable applications have arisen: ranging from Galois theory, generation of simple groups and density of derangement graphs. Each of these applications has a number of interesting open questions and conjectures, which might benefit from having a well-established theory of weak normal coverings. Therefore, it is important to collect all partial results already available in the literature and to have a comprehensive analysis of the most basic case, that is, weak normal 2-coverings of non-abelian simple groups. This detailed information might shed new light in some of these applications.

Most of our work is concerned with simple classical groups and on their conjugacy classes. In some occasions, our results on conjugacy classes are more general than our needs. We hope that these more general results can be used to shape a general theory of weak normal coverings of non-abelian simple groups.

Firenze, Italy
Milano, Italy
Milano, Italy

Daniela Bubboloni
Pablo Spiga
Thomas Stefan Weigel

Acknowledgement

We wish to express our sincere gratitude to Andrea Lucchini and to Gunter Malle for reading a preliminary draft of this manuscript. We also thank the anonymous referees for their helpful comments.

Contents

1	**Introduction**	1
	1.1 The Weak Normal Covering Number and the Normal Covering Number of the Finite Non-abelian Simple Groups	3
	1.2 The Normal 2-Coverings of the Almost Simple Groups and the Weak Normal Covering Number of the Almost Simple Groups	8
	1.3 The Invariably Generating Graph and the Aut-invariably Generating Graph	10
	1.4 The Erdős–Ko–Rado Theorem and the Derangement Graph	13
	1.5 Further on Derangements: The Boston-Shalev Conjecture	16
	1.6 Normal 2-Coverings for Arbitrary Finite Groups	17
	1.7 Normal Coverings and Kronecker Classes	20
	1.8 Structure of the Book and Comments	22
2	**Preliminaries**	27
	2.1 Classical Groups	27
	2.2 Normal and Weak Normal Coverings of Classical and Simple Classical Groups	28
	2.3 From Weak Normal Coverings to Normal Coverings	29
	2.4 Action of $\hat{G}$ on Its Natural Module	30
	2.5 Huppert's Theorem and Singer Cycles	31
	2.6 Some Arithmetics	33
	2.7 Primitive Prime Divisors	35
	2.8 Further Facts on Singer Cycles	36
	2.9 The ppd-Elements	38
	2.10 Bertrand Elements	40
	2.11 The Spinor Norm and the Bertrand Elements	41
3	**Linear Groups**	45

4	**Unitary Groups**	51
	4.1 Small Dimensional Unitary Groups	52
	4.2 Large Dimensional Unitary Groups	61
5	**Symplectic Groups**	65
	5.1 Small Dimensional Symplectic Groups	65
	5.2 Large Dimensional Symplectic Groups	80
6	**Odd Dimensional Orthogonal Groups**	87
	6.1 Small Odd Dimensional Orthogonal Groups	88
	6.2 Large Odd Dimensional Orthogonal Groups	90
	6.3 An Auxiliary Result	96
7	**Orthogonal Groups with Witt Defect 1**	101
	7.1 Small Even Dimensional Orthogonal Groups of Witt Defect 1	103
	7.2 Large Even Dimensional Orthogonal Groups of Witt Defect 1	107
8	**Orthogonal Groups with Witt Defect 0**	113
	8.1 Orthogonal Groups of Witt Defect 0 of Dimension at Least 10	114
	8.2 Eight Dimensional Orthogonal Groups with Witt Defect 0	118
9	**Proofs of the Main Theorems**	125
10	**Almost Simple Groups Having Socle a Sporadic Simple Group**	129
11	**Dropping the Maximality**	135
	11.1 Linear Groups: Cases (3)–(7)	141
	11.2 Unitary Groups: Cases (8) and (9)	148
	11.3 Exceptional Groups: Case (2)	151
	11.4 Exceptional Groups: Case (1)	152
	11.5 Symplectic Groups: Cases (10) and (11)	153
	11.6 Proof of Theorem 2 of Burness and Tong-Viet	158
12	**Degenerate Normal 2-Coverings**	161
	12.1 The Degenerate Normal 2-Coverings of the First Type	161
	12.2 The Degenerate Normal 2-Coverings of the Second Type	162
References	....	175

Chapter 1
Introduction

It is well known that a finite group cannot be the union of conjugates of a proper subgroup. Indeed, let G be a finite group and let H be a proper subgroup of G. Then

$$\left|\bigcup_{g\in G}(H \setminus \{1\})^g\right| \leq |H \setminus \{1\}||G : H| = (|H| - 1)|G : H|$$
$$= |G| - |G : H| < |G| - 1,$$

because H is properly contained in G. Thus some element of G is conjugate to none of the elements of H.[1]

However, there are examples of finite groups which are the union of conjugates of two proper subgroups. For instance, the symmetric group Sym(3) of degree 3 is the union of its normal subgroup Alt(3) and of its Sylow 2-subgroups. There are more intriguing examples of this phenomenon. For instance, it was already shown by Dye [33] in 1979 that, the symplectic group $\mathrm{Sp}_n(q)$ is the union of conjugates of its subgroups $\mathrm{SO}_n^-(q)$ and $\mathrm{SO}_n^+(q)$, when q is even.

Definition 1.1 Let k be a positive integer and let G be a finite non-cyclic group. A ***normal k-covering*** of G is a set $\mu = \{H_1, \ldots, H_k\}$ of k proper subgroups of G with the property that every element of G belongs to the conjugate H_i^g, for some $i \in \{1, \ldots, k\}$ and for some $g \in G$, that is,

$$G = \bigcup_{i=1}^{k} \bigcup_{g \in G} H_i^g.$$

---

[1] Higman et al. [58] constructed an infinite group G with the property that all the non-identity elements of G are conjugate. Clearly, G is simple.

© The Author(s), under exclusive license to Springer Nature Switzerland AG 2024
D. Bubboloni et al., *Normal 2-Coverings of the Finite Simple Groups and their Generalizations*, Lecture Notes in Mathematics 2352,
https://doi.org/10.1007/978-3-031-62348-6_1

We refer to $H_1, \ldots, H_k$ as the **components** of μ. If $H_1, \ldots, H_k$ are maximal subgroups of G, we refer to them as **maximal components**. Clearly, if G is a cyclic group, then G admits no normal k-covering, because the generators of G lie in no proper subgroup.

The **normal covering number** of the group G, denoted by $\gamma(G)$, is the smallest integer k such that G admits a normal k-covering. Note that in a normal k-covering $\{H_1, \ldots, H_k\}$ with $k = \gamma(G)$, the proper subgroups $H_1, \ldots, H_k$ are in distinct G-conjugacy classes.

Coming back to our examples above, from this definition, we may write $\gamma(\mathrm{Sym}(3)) = 2$ and, when q is even, $\gamma(\mathrm{Sp}_n(q)) = 2$.

Finite groups having normal covering number 2 are often algebraically and combinatorially very interesting; for instance, the examples of Dye have been used in [50] to give new solutions to Perlis' equation [78] in algebraic number fields. One of the first motivations for investigating finite groups having normal covering number 2 goes back to a problem in Galois theory (for more details see [8, Section 1]) and is linked to the study of intersective polynomials, that is, integer polynomials having a root modulo p, for every prime number p (see [15] and [83]). Recently, simple groups having normal covering number 2 have been used to construct sparsely connected invariably generating graphs [44]. There is also an unexpected connection between the theory of transitive permutation groups G in which every derangement is a p-element for some prime p and the normal 2-coverings of G. Indeed, as observed in [21], G has such property if and only if G admits a normal 2-covering with components given by a point stabilizer and a Sylow p-subgroup of G. As an application of our work, we give a brief proof of one of the main results in [21] in Sect. 11.6.

We also recall that the normal covering number $\gamma(G)$ has connections with some questions about the generation of the group G. Consider the number $\kappa(G)$, introduced by Britnell and Maróti [12] as the maximum size of a set X of conjugacy classes of G such that any pair of elements from distinct classes in X generates G. As already observed in [17], from this definition, it is clear that

$$\kappa(G) \leq \gamma(G). \tag{1.1}$$

Now, Guralnick and Malle [52, Theorem 1.3] and Kantor et al. [63, Theorem 1.3] have shown that for non-abelian finite simple groups we have

$$\kappa(G) \geq 2. \tag{1.2}$$

In this monograph we determine, among other things, the finite simple groups G having normal covering number as small as possible, that is, with $\gamma(G) = 2$. Thus, in view of (1.1) and (1.2), our results also allow to describe some of the finite simple groups G attaining the minimum value for κ, that is $\kappa(G) = 2$.

Actually, in applications, it is often important to consider a more general situation. This more general situation arises for instance when investigating small cliques in derangement graphs of permutation groups [76] and also in the reduction theorem in [42] for investigating arbitrary finite groups having small normal covering number, a similar reduction theorem appears also in [82]. Another typical example where this more general situation is important is the study of Kronecker classes in number fields, see [79–82, 85].

Definition 1.2 Let k be a positive integer and let G be a finite non-cyclic group. A *weak normal k-covering* of G is a family $\mu = \{H_1, \ldots, H_k\}$ of k distinct proper subgroups of G with the property that every element of G belongs to H_i^g, for some $i \in \{1, \ldots, k\}$ and for some $g \in \mathrm{Aut}(G)$, that is,

$$G = \bigcup_{i=1}^{k} \bigcup_{g \in \mathrm{Aut}(G)} H_i^g.$$

As for normal coverings, we refer to $H_1, \ldots, H_k$ as the *components* of the weak normal k-covering and if $H_1, \ldots, H_k$ are maximal subgroups of G, we refer to them as *maximal components*. The *weak normal covering number* of G, denoted by $\gamma_w(G)$, is the smallest integer k such that G admits a weak normal k-covering. Since $\mathrm{Aut}(G)$ contains all the inner automorphisms of G, we have $\gamma_w(G) \leq \gamma(G)$.

Note that in a weak normal k-covering $\{H_1, \ldots, H_k\}$ with $k = \gamma_w(G)$, the proper subgroups $H_1, \ldots, H_k$ are in distinct $\mathrm{Aut}(G)$-conjugacy classes.

Observe that, if G is a group having (weak) normal covering number k, then G has a (weak) normal k-covering whose components are maximal subgroups of G. In this monograph we are mainly concerned with maximal components.

For instance, $\gamma_w(\mathrm{Sym}(3)) = 2$ and, for every non-cyclic elementary abelian p-group G, we have $\gamma_w(G) = 1$.

In the remainder of this chapter, we begin by presenting our key findings, followed by a comprehensive discussion on the practical applications and underlying motivations of our work, which were only briefly touched upon in this introduction.

1.1 The Weak Normal Covering Number and the Normal Covering Number of the Finite Non-abelian Simple Groups

The following is the main result of the monograph.

Theorem 1.3 *Let G be a finite non-abelian simple group. Then $\gamma_w(G) \geq 2$. Moreover,*

(i) $\gamma_w(G) = 2$ *if and only if G appears in the first column of* Tables 1.3, 1.4, 1.5, 1.6, and 1.7;

(ii) $\gamma(G) = 2$ *if and only if in* Tables 1.3, 1.4, 1.5, 1.6, and 1.7 *the group G appears in the first column and at least one number is different from 0 in the fifth column.*

Prior to delving into the detailed examination of the groups emerging from Theorem 1.3, we briefly divert our attention to earlier studies, practical applications, and unresolved inquiries, with further discussion on these topics later in the chapter. It is important to highlight at this juncture that our primary finding furnishes a comprehensive and meticulous account of non-abelian simple groups with a (weak) covering number of 2.

Saxl [85] in his investigation on Kronecker classes has proved that $\gamma_w(G) \geq 2$, for every finite non-abelian simple group. However, since his main concern was showing $\gamma_w(G) \neq 1$, his work does not attempt to classify the finite non-abelian simple groups attaining the minimum $\gamma_w(G) = 2$. We do this in our main result Theorem 1.3. We emphasize that the inequality $\gamma_w(G) \geq 2$ can also be deduced from the results in [63] and [52], as commented in Sect. 1.3. The challenging aspect (and a focal point of this monograph) of Theorem 1.3 relies on the classification of the finite non-abelian simple groups for which the functions γ_w and γ attain their minimum value 2.

Weak normal 1-coverings of finite groups have already appeared in the literature a few times in the study of Kronecker classes [79–82], see Sect. 1.7. It is still open an interesting question of Neumann and Praeger [65, Problem 11.71]. We report here, to give further evidence, their question. Let A be a finite group with a normal subgroup G. A subgroup H of G is called an A-**covering subgroup** of G if

$$G = \bigcup_{a \in A} H^a.$$

Clearly, if H is an A-covering subgroup of G, then $\gamma_w(G) = 1$ and $\{H\}$ is a weak normal 1-covering of G.

Question Is there a function $f : \mathbb{N} \to \mathbb{N}$ such that whenever $H < G < A$, where A is a finite group, G is a normal subgroup of A of index n, and H is an A-covering subgroup of G, we have $|G : H| \leq f(n)$?

We will now introduce key terminology to facilitate understanding the statement of Theorem 1.3.

Notation 1.4 We explain the notation in Tables 1.3, 1.4, 1.5, 1.6, and 1.7. There are six columns in total. In the first column, we designate the simple group G. In the sixth column, we have reported the number of weak normal 2-coverings of G with maximal components up to $\mathrm{Aut}(G)$-*conjugacy*, where we say that two weak normal 2-coverings having components $\{H_1, K_1\}$ and $\{H_2, K_2\}$ are $\mathrm{Aut}(G)$-*conjugate* if there exist $\varphi, \psi \in \mathrm{Aut}(G)$ with either

$$H_1^\varphi = H_2 \text{ and } K_1^\psi = K_2, \text{ or } H_1^\varphi = K_2 \text{ and } K_1^\psi = H_2.$$

In the second and in the third column, we present the maximal components H and K for the weak normal 2-coverings of G. We present a row for each $\mathrm{Aut}(G)$-conjugacy class. In particular, for a fixed G, the number of rows appearing in

Tables 1.3, 1.4, 1.5, 1.6, and 1.7 for G is exactly the number appearing in the sixth column. For instance, when $G := A_6$, there are two $\text{Aut}(G)$-conjugacy classes of weak normal 2-coverings and in the two rows concerning A_6 we have reported the two weak normal 2-coverings with maximal components $\{A_6 \cap (S_2 \times S_4), A_5\}$ and $\{A_6 \cap (S_3 \text{wr} S_2), A_5\}$.

In the fourth column, we collect some very basic comments (typically recalling some isomorphisms between the groups in the list).

We finally explain the number appearing in the fifth column. We say that two weak normal 2-coverings of G having components $\{H_1, K_1\}$ and $\{H_2, K_2\}$ are *G-conjugate* if there exist $\varphi, \psi \in G$ with either

$$H_1^\varphi = H_2 \text{ and } K_1^\psi = K_2, \text{ or } H_1^\varphi = K_2 \text{ and } K_1^\psi = H_2.$$

Clearly, this defines an equivalence relation on the set of weak normal 2-coverings, which refines the equivalence relation given by the $\text{Aut}(G)$-conjugacy classes.

Now, let $\{H, K\}$ be a weak normal 2-covering of G with maximal components appearing in some row of Tables 1.3, 1.4, 1.5, 1.6, and 1.7. The $\text{Aut}(G)$-conjugacy class

$$\{\{H^\varphi, K^\psi\} \mid \varphi, \psi \in \text{Aut}(G)\}$$

is a union of G-conjugacy classes. Observe that only some of these G-conjugacy classes (possibly none) give rise in fact to normal 2-coverings and, in the fifth column we report their number.

Returning to our example of $G := A_6$, the $\text{Aut}(G)$-conjugacy class represented by the weak normal 2-covering with components $\{A_6 \cap (S_2 \times S_4), A_5\}$ splits into exactly two G-conjugacy classes of normal 2-coverings. This inference is drawn from Table 1.3, where the fifth column corresponding to this weak normal 2-covering displays a 2.

A similar comment applies for the components $\{A_6 \cap (S_3 \text{wr} S_2), A_5\}$. Analogously, from Table 1.3, $G := G_2(3)$ has two $\text{Aut}(G)$-conjugacy classes of weak normal 2-coverings, but none of the corresponding G-classes gives rise to a normal 2-covering. Indeed, as above, this inference is drawn from Table 1.3, where the fifth column corresponding to the weak normal 2-coverings of $G_2(3)$ displays a 0.

When an $\text{Aut}(G)$-conjugacy class gives rise to at least two normal 2-coverings with maximal components, information about the embeddings of the corresponding components in G can be found directly in the statements related to G. For instance, the two normal 2-coverings of $\text{Sp}_4(q)$, for $q \geq 8$ even, arising from the unique $\text{Aut}(\text{Sp}_4(q))$-conjugacy class of weak normal 2-coverings are described in Lemma 5.4.

Note that, by adding the numbers appearing in the fifth column for a given group G, we obtain the number of normal 2-coverings of G, up to G-conjugacy.

Our exhaustive proof of Theorem 1.3 not only allows the classification of non-abelian simple groups G with a weak normal 2-covering but also provides a detailed classification of the constituent components $\{H, K\}$ within such coverings, when both H and K are maximal subgroups of G. This endeavor heavily relies on the thorough classification of the maximal subgroups within non-abelian simple groups. Our analysis focuses on coverings up to G-conjugacy and $\mathrm{Aut}(G)$-conjugacy, resulting in the following refinement.

Theorem 1.5 *Let G be a finite non-abelian simple group and let $\mu = \{H, K\}$ be a weak normal 2-covering of G with maximal components. Then the pair (H, K) appears in Tables 1.3, 1.4, 1.5, 1.6, and 1.7, up to $\mathrm{Aut}(G)$-conjugacy. Moreover, μ gives rise to at least a normal 2-covering of G if and only if in the corresponding row of the fifth column of those tables appears a number greater than 0.*

The notation we use to describe the components H and K is standard. In particular, for the simple classical groups we use the notation in [9, 66]. We try to give as much information as possible on H and K in Tables 1.3, 1.4, 1.5, 1.6, and 1.7, however when this is not possible we just give a rough description.

In particular, Theorem 1.5 gives a complete classification of the finite simple groups G with $\gamma(G) = 2$ or with $\gamma_w(G) = 2$ and of the (weak) normal 2-coverings with maximal components. An immediate consequence of Theorem 1.5 is the following corollary.

Corollary 1.6 *Let G be a finite non-abelian simple group. Then $\gamma_w(G) = 2$ if and only if one of the following holds*

(1) $\gamma(G) = 2$ and G is isomorphic to one of the groups listed in Table 1.1,
(2) G is the alternating group A_9,
(3) G is the sporadic group M_{12},
(4) G is the exceptional group of Lie type $G_2(3)$,
(5) G is the unitary group $\mathrm{PSU}_6(2)$,
(6) G is the orthogonal group $\mathrm{P}\Omega_8^+(2)$ or $\mathrm{P}\Omega_8^+(3)$.

In Table 1.1, we have highlighted some isomorphisms among non-abelian simple groups. Recall also that $\mathrm{P}\Omega_4^+(q) \cong \mathrm{PSL}_2(q) \times \mathrm{PSL}_2(q)$ is not simple.

Corollary 1.6 finds its greatest utility when coupled with Theorem 1.3. For example, according to Corollary 1.6, $\gamma_w(\mathrm{P}\Omega_8^+(3))$ and $\gamma(\mathrm{P}\Omega_8^+(3)) > 2$. Should we require information on the maximal components of a weak normal 2-covering of $\mathrm{P}\Omega_8^+(3)$, we can refer to Theorem 1.3, particularly Table 1.7. This table reveals that $\mathrm{P}\Omega_8^+(3)$ exhibits four distinct weak normal 2-coverings with maximal components. However, upon consulting the fifth column, it becomes evident that none of these produce a normal 2-covering.

It follows immediately from Corollary 1.6 that, apart from six exceptional cases, for each non-abelian simple group G, we have $\gamma(G) = 2$ if and only if $\gamma_w(G) = 2$. We actually conjecture that a similar pattern holds in general and hence, in this case, nature is not as diverse as it can possibly be: $\gamma(G)$ always coincides with $\gamma_w(G)$, except for a "low level noise".

1.1 The (Weak) Normal Covering Number of Non-abelian Simple Groups

Table 1.1 Classification of finite simple groups G with $\gamma(G) = 2$

Type	Groups
Alternating	A_5, A_6, A_7, A_8
Sporadic	M_{11}
Exceptional	$G_2(2^a)$ with $a \geq 2$, $G_2(2)'$, $^2G_2(3)'$
	$^2F_4(2)'$, $F_4(3^a)$ with $a \geq 1$
Linear	$\mathrm{PSL}_2(q)$ with $q \geq 4$
	$\mathrm{PSL}_3(q)$
	$\mathrm{PSL}_4(q)$
Unitary	$\mathrm{PSU}_2(q) \cong \mathrm{PSL}_2(q)$ with $q \geq 4$
	$\mathrm{PSU}_3(3^a)$ with $a \geq 1$, $\mathrm{PSU}_3(5)$
	$\mathrm{PSU}_4(q)$
Symplectic	$\mathrm{PSp}_2(q) \cong \mathrm{PSL}_2(q)$ with $q \geq 4$
	$\mathrm{PSp}_4(3) \cong \mathrm{PSU}_4(2)$, $\mathrm{PSp}_4(2)' \cong A_6$
	$\mathrm{PSp}_n(2^a)$ with $n \geq 4$, $a \geq 1$ and $(n,a) \neq (4,2)$
	$\mathrm{PSp}_6(3^a)$ with $a \geq 1$
Orthogonal	$\Omega_3(q) \cong \mathrm{PSL}_2(q)$, $\Omega_5(3) \cong \mathrm{PSp}_4(3)$,
	$\mathrm{P}\Omega_4^-(q) \cong \mathrm{PSL}_2(q^2)$, $\mathrm{P}\Omega_6^-(q) \cong \mathrm{PSU}_4(q)$
	$\mathrm{P}\Omega_6^+(q) \cong \mathrm{PSL}_4(q)$

Conjecture 1.7 There exists a function $f : \mathbb{N} \to \mathbb{N}$ such that, if G is a finite non-abelian simple group with $\gamma_w(G) = m$, then either $\gamma(G) = m$, or $|G| \leq f(m)$, or $G \cong \mathrm{P}\Omega_n^+(q)$.

Clearly, for each m, it is of independent interest to classify the finite non-abelian simple groups with $\gamma_w(G) = m \neq \gamma(G)$. Note that Corollary 1.6 gives that classification when $m = 2$. In Conjecture 1.7, we believe that $\gamma_w(\mathrm{P}\Omega_n^+(q))$ and $\gamma(\mathrm{P}\Omega_n^+(q))$ differ for infinitely many values of n and q, because of the role of graph automorphisms.

Inspired by the work in this monograph, we now also make a conjecture concerning normal coverings of finite non-abelian simple groups of bounded cardinality. Broadly speaking, we conjecture that, for every positive integer c, the number of finite simple groups having normal covering number at most c is "small". However, we need to clarify what we mean by small.

Conjecture 1.8 There exists a function $f : \mathbb{N} \to \mathbb{N}$ such that if G is a finite non-abelian simple group with $\gamma(G) \leq c$, then one of the following holds:

- $|G| \leq f(c)$,
- G is a finite simple group of Lie type having Lie rank at most $f(c)$,
- $G = \mathrm{Sp}_n(q)$ with q even.

We have little evidence towards Conjecture 1.8. For linear groups, it follows from the work of Britnell and Maróti [12]. For alternating groups, it follows from the linear bounds on the normal covering number of the symmetric and alternating groups in [17, 18].

Clearly, for a fixed c, the classification of the finite non-abelian simple groups having normal covering number at most c can be rather hard and, in this context, our monograph deals with the first meaningful case $c = 2$.

Our Tables 1.3, 1.4, 1.5, 1.6, and 1.7 look rather complex and difficult and, in a sense, they are, because we are pinning down one by one all weak normal 2-coverings. However, at a careful analysis, some pattern arises for classical groups. Indeed, for the classical groups, there are weak normal 2-coverings which hold for each value of q: these might be regarded as generic weak normal 2-coverings, as defined in [16]. All other weak normal 2-coverings that arise are when q is small.

The reader might notice some similarities between the maximal components appearing in Tables 1.3, 1.4, 1.5, 1.6, and 1.7 and the non-trivial maximal factorizations of simple groups classified by Liebeck et al. in [71, 72]. Simple groups, such as the symplectic groups $\mathrm{PSp}_6(3^f)$ with $f \geq 2$, admit normal 2-coverings, but lack non-trivial factorizations. Conversely, other simple groups, such as alternating groups of prime degree $p \geq 11$, have non-trivial factorizations, but lack normal 2-coverings. Hence, there exists no direct correlation between normal 2-coverings and factorizations. Similarly, the reader might notice some similarities between the groups appearing in Tables 1.3, 1.4, 1.5, 1.6, and 1.7 and the factorizations of the form $G = \mathbf{N}_G(\langle x \rangle)\mathbf{N}_G(\langle y \rangle)$ of the almost simple groups G obtained in [47].

1.2 The Normal 2-Coverings of the Almost Simple Groups and the Weak Normal Covering Number of the Almost Simple Groups

Weak normal 2-coverings of non-abelian simple groups allow us to shed light on the normal 2-coverings of almost simple groups. In this section, we elucidate how Theorem 1.3 has enabled us to gather information on almost simple groups with a normal covering number of 2.

Let A be an *almost simple* group with socle the non-abelian simple group G. We recall that this means that G is the unique minimal normal subgroup of A. Suppose A admits a normal 2-covering with components $H < A$ and $K < A$. There are only three kinds of choice for those components:

ND: $H \not\geq G$ and $K \not\geq G$,
DI: $G \leq H \cap K$,
DII: replacing H with K if necessary, $H \geq G$ and $K \not\geq G$.

The label ND stands for non-degenerate, DI for degenerate of first type and DII for degenerate of second type.

In the first case, that is ND, from $A = \bigcup_{a \in A}(H^a \cup K^a)$, we deduce

$$G = \bigcup_{a \in A}((H \cap G)^a \cup (K \cap G)^a)$$

1.2 The (Weak) Normal 2-Coverings of the Almost Simple Groups

with $H \cap G$ and $K \cap G$ proper subgroups of G. Therefore $\gamma_w(G) = 2$ and hence G is one of the groups in the first column of Tables 1.3, 1.4, 1.5, 1.6, and 1.7. Moreover, building on that idea, we prove the following result on the weak normal covering number of almost simple groups.

Theorem 1.9 *Let X be a finite almost simple group. Then one of the following holds*

(1) $\gamma_w(X) \geq 2$,
(2) *the socle G of X is $P\Omega_8^+(q)$, q is odd, $PO_8^+(q) \leq X$, $\mathrm{Aut}(X)$ contains a triality automorphism of G and X does not contain a triality automorphism of G. Moreover, each weak normal 1-covering $\{H\}$ of X satisfies $G \leq H$.*

We observe that there exist almost simple groups X with $\gamma_w(X) = 1$ and hence instances in part (2) of Theorem 1.9 do arise. For instance, let $G := P\Omega_8^+(q)$ with q an odd prime number and let $A := \mathrm{Aut}(G)$. Observe that the group of outer automorphisms $\mathrm{Out}(G)$ of G is isomorphic to the symmetric group S_4. Therefore, A contains a normal subgroup X with $G \leq X$ and with X/G equal to the Klein group of $\mathrm{Out}(G)$. Let $H := PO_8^+(q)$ and observe that

$$G \leq H \leq X$$

and $|H : G| = |X : H| = 2$. Now, the conjugates of H under A cover the whole of X, that is, $X = \bigcup_{a \in A} H^a$, and hence $\{H\}$ is a weak normal 2-covering of X. The main contribution in Theorem 1.9 is that all almost simple groups X with $\gamma_w(X) = 1$ arise from a similar construction. For more details on the possible groups X arising in part (2) of Theorem 1.9 see the proof of Theorem 1.9 and Remark 9.3 in Chap. 9.

Degenerate normal 2-coverings of A of the first type correspond to normal 2-coverings of the quotient group A/G. Note that, when $\mathrm{Aut}(G)/G$ is cyclic (like for sporadic simple groups and alternating groups of degree different from 6), these normal 2-coverings do not arise. In Theorem 12.3, we do give a complete classification of the non-abelian simple groups admitting a 2-covering of this type.

The degenerate normal 2-coverings of A of the second type are studied in Theorem 12.5. The analysis of these degenerate normal 2-coverings is intimately related to the Memoir of Guralnick et al. [54] and to the theory of the Shintani descent. While the specifics of Theorem 12.5 are too intricate for this introductory section, we offer a straightforward example of such peculiar normal 2-covering. The group $A := P\Gamma L_2(8)$ admits a normal 2-covering using $H := PSL_2(8)$ and $K := \mathbf{N}_A(P)$, where P is a 3-Sylow of A. This illustration belongs to a larger collection of examples outlined in Theorem 12.5, (2b). Despite our efforts, a comprehensive classification of normal 2-coverings of the second type for A eludes us, due to inherent challenges in achieving a general result in this domain. Consider A as an almost simple group with socle G, and let H, K be a normal 2-covering of A of the second type, where $G \leq H$ and $G \not\leq K$. In Theorem 12.5, we provide insights into G, H, and K only when A/G is nilpotent. Furthermore, while we cannot fully describe H and K, we do offer information on the maximal subgroups of A containing them, respectively. A notable caveat in Theorem 12.5 is our inability, as of now, to ascertain whether all enumerated cases indeed yield degenerate normal 2-coverings of the second type. For a comprehensive examination, please refer to Chap. 12.

We will now delve into further details regarding the applications and motivations behind our work.

1.3 The Invariably Generating Graph and the Aut-invariably Generating Graph

The connection between $\gamma(G)$ and $\kappa(G)$ runs deeper than what we have mentioned in (1.1). The right context to talk about this is the *invariably generating graph*. Let G be a group and let $g_1, \ldots, g_\kappa \in G$. Then, $\{g_1, \ldots, g_\kappa\}$ is said to *invariably generate* G if, for every $x_1, \ldots, x_\kappa \in G$, we have

$$G = \langle g_1^{x_1}, \ldots, g_\kappa^{x_\kappa} \rangle.$$

The reference work on invariable generation is the paper of Kantor et al. [63], where among other things it is proved [63, Theorem 1.3] that every non-abelian finite simple group is invariably generated by two elements. Recall that there is a combinatorial gadget $\Lambda(G)$, the *invariably generating graph* of G, that can be efficiently used to investigate invariably generating pairs. The vertices of $\Lambda(G)$ are the non-trivial conjugacy classes of G and two conjugacy classes g_1^G and g_2^G are declared to be adjacent if the set $\{g_1, g_2\}$ invariably generates G. Using this terminology, $\kappa(G)$ is the clique number of $\Lambda(G)$, that is, the largest cardinality of a set of pair-wise adjacent vertices of $\Lambda(G)$. The invariably generation of groups has tight connection with groups admitting Beauville structures and with the spread of a group. It is an interesting invariant associated to a group that has recently received some attention [43, 45, 46] and its investigation requires some deep results, like the Fulman-Guralnick solution [37–40] of the Boston-Shalev [7] conjecture on derangements in finite simple groups.

Here, we simply observe that, at the moment, for non-abelian simple groups, we only have a handful of examples where $\gamma(G) \neq \kappa(G)$. For instance, $\kappa(G_2(3)) = 2 = \kappa(J_2)$, but $\gamma(G_2(3)) \neq 2 \neq \gamma(J_2)$. We are therefore wondering whether there exist only finitely many non-abelian simple groups G with $\kappa(G) \neq \gamma(G)$. We doubt very much that this is the case and hence we propose the following more general question.

Problem 1.10 Classify the finite non-abelian simple groups G with $\kappa(G) \neq \gamma(G)$.

In the monograph, we completely classify the sporadic simple groups G with $\kappa(G) \neq \gamma(G)$ (see Table 1.2). Moreover, motivated by Theorem 1.3, by (1.1) and by (1.2), we propose the following.

Problem 1.11 Classify the finite non-abelian simple groups G with $\kappa(G) = 2$.

1.3 The (Aut-)invariably Generating Graph

Table 1.2 Values of $\gamma, \kappa, \gamma_w, \kappa_w$ for the almost simple groups with sporadic socle. Observe that in all cases $\gamma - \kappa_w \leq 1$

Grp	$\gamma(\cdot)$	$\gamma_w(\cdot)$	$\kappa(\cdot) = \kappa_w(\cdot)$	Grp	$\gamma(\cdot) = \gamma_w(\cdot)$	$\kappa(\cdot) = \kappa_w(\cdot)$
M_{11}	2	2	2	$M_{12}.2$	3	2
M_{12}	3	2	2	$M_{22}.2$	3	2
M_{22}	3	3	3	$J_2.2$	3	3
M_{23}	3	3	3	$Suz.2$	4	4
M_{24}	3	3	3	$HS.2$	3	3
J_1	4	4	4	$McL.2$	3	2
J_2	3	3	2	$He.2$	4	3
J_3	3	3	3	$Fi_{22}.2$	4	3
J_4	7	7	7	Fi_{24}	5	5
Co_1	5	5	4	$HN.2$	4	4
Co_2	4	4	3	$O'N.2$	3	3
Co_3	3	3	3	$J_3.2$	3	3
Fi_{22}	4	4	3			
Fi_{23}	5	5	5			
Fi'_{24}	5	5	5			
Suz	4	4	4			
Ru	3	3	3			
Ly	5	5	5			
$O'N$	3	3	3			
McL	3	3	3			
HS	3	3	2			
HN	4	4	4			
He	3	3	3			
Th	5	5	4			
$F_{2+} = B$	7	7	7			
$F_1 = M$	9	9	9			

Inspired by our definition of weak normal covering and by some results in the recent literature like Theorem 5.1 in [63] and Corollary 7.2 in [52], we propose a new graph to be associated with a group. Let G be a group and let $g_1, \ldots, g_\kappa \in G$. Then, $\{g_1, \ldots, g_\kappa\}$ is said to **Aut-invariably generate** G if, for every $x_1, \ldots, x_\kappa \in \mathrm{Aut}(G)$, we have

$$G = \langle g_1^{x_1}, \ldots, g_\kappa^{x_\kappa} \rangle.$$

Of course, if $\{g_1, \ldots, g_\kappa\}$ Aut-invariably generates G, then $\{g_1, \ldots, g_\kappa\}$ invariably generates G too. We define the **Aut-invariably generating graph** $\Upsilon(G)$, considering as vertices the non-trivial $\mathrm{Aut}(G)$-conjugacy classes of G and two classes $g_1^{\mathrm{Aut}(G)}$ and $g_2^{\mathrm{Aut}(G)}$ to be adjacent if the set $\{g_1, g_2\}$ Aut-invariably generates G. We define next, $\kappa_w(G)$ as the clique number of $\Upsilon(G)$. It is easily checked that the following inequalities hold

$$\kappa_w(G) \leq \gamma_w(G), \qquad \kappa_w(G) \leq \kappa(G). \tag{1.3}$$

By [63, Theorem 5.1] or by [52, Corollary 7.2], for every non-abelian simple group G with the exception of $G = P\Omega_8^+(q)$ for $q \leq 3$, we have

$$\kappa_w(G) \geq 2. \tag{1.4}$$

Using the computer algebra systems GAP [89] and magma [6] we computed the values of κ_w and γ_w for the two missing groups above:

$$\kappa_w(P\Omega_8^+(2)) = 1 \quad \text{and} \quad \kappa_w(P\Omega_8^+(3)) = \gamma_w(P\Omega_8^+(3)) = 2. \tag{1.5}$$

In particular, by (1.3) and (1.4), we deduce that, for every non-abelian simple group G with the exception of $G = P\Omega_8^+(q)$ for $q = 2$, we have $\gamma_w(G) \geq 2$. On the other hand, in Lemma 8.4, we show that $\gamma_w(P\Omega_8^+(2)) = 2$ (see also Table 1.7). Thus, with no exceptions, for every non-abelian simple group G, we have

$$\gamma_w(G) \geq 2. \tag{1.6}$$

In Table 1.2 we have reported the values of γ, γ_w, κ and κ_w for the almost simple groups having socle a sporadic group. These values are computed by the invaluable help of computer algebra systems GAP [89] and magma [6]. Details on these computations are in Chap. 10.

Looking at Table 1.2 we see examples of sporadic simple groups G with $\gamma_w(G) \neq \kappa_w(G)$. For instance, $\kappa_w(J_2) = 2 \neq \gamma_w(J_2) = 3$. On the other hand there are no examples of sporadic simple groups G with $\kappa_w(G) \neq \kappa(G)$. We then propose the following question.

Problem 1.12 Classify the finite non-abelian simple groups G with $\kappa_w(G) \neq \gamma_w(G)$ and those with $\kappa_w(G) \neq \kappa(G)$.

Of course Table 1.2 completely classifies the sporadic simple groups G with $\kappa_w(G) \neq \gamma_w(G)$.

Moreover, motivated by Theorem 1.3, by (1.3), (1.4) and (1.5), we propose the following last problem.

Problem 1.13 Classify the finite non-abelian simple groups G with $\kappa_w(G) = 2$.

Certainly, by (1.3), some of the finite non-abelian simple groups G such that $\kappa_w(G) = 2$ come as those satisfying $\gamma_w(G) = 2$ and hence are described in Theorem 1.3.

1.4 The Erdős–Ko–Rado Theorem and the Derangement Graph

Very recently, weak normal coverings have proved useful in investigations on the density of derangement graphs. It seems plausible that these fruitful applications can answer further open questions in the area.

One of the most beautiful results in extremal combinatorics is the Erdős–Ko–Rado theorem: let n and k be positive integers with $1 \le 2k < n$ and let $\mathcal{F}$ be a family of k-subsets of $\{1, \ldots, n\}$. If any two elements from $\mathcal{F}$ intersect in at least one point, then $|\mathcal{F}| \le \binom{n-1}{k-1}$. Moreover, the inequality is attained if and only if there exists $x \in \{1, \ldots, n\}$ such that each element from $\mathcal{F}$ contains x.

There are various analogues of the Erdős–Ko–Rado theorem for a number of combinatorial structures. Here we are interested in the analogue for permutation groups. Let G be a finite permutation group on Ω. A subset $\mathcal{F}$ of G is said to be **intersecting** if, for any two elements $g, h \in \mathcal{F}$, gh^{-1} fixes some point of Ω. This is a very natural definition; indeed, by writing g as the n-tuple $(1^g, 2^g, \ldots, n^g)$, we see that gh^{-1} fixes some point of Ω if and only if the n-tuples corresponding to g and h agree in at least one coordinate. Therefore, somehow, this mimics the definition of intersecting sets in the original Erdős–Ko–Rado theorem.

Observe that, for every $\omega \in \Omega$, the point stabilizer G_ω is intersecting. More generally, each coset of the stabilizer of a point is an intersecting set. Unfortunately, only rarely, G_ω is an intersecting set of maximal size in G and hence no analogue of the Erdős–Ko–Rado theorem holds for arbitrary permutation groups.[2] Even when $|G_\omega|$ is the maximal cardinality of an intersecting set for G, it is far from being true that all intersecting sets attaining the bound $|G_\omega|$ are cosets of the stabilizer of a point.[3] These two difficulties make investigations on intersecting sets of maximal size in arbitrary permutation groups more interesting and challenging.

Let $\omega \in \Omega$ with G_ω having maximum cardinality among point stabilizers.[4] The **intersection density** of the intersecting family $\mathcal{F}$ of G is defined by

$$\rho(\mathcal{F}) = \frac{|\mathcal{F}|}{|G_\omega|}.$$

---

[2] For instance, if we let the alternating group Alt(5) act on the ten 2-subsets of $\{1, 2, 3, 4, 5\}$, we see that Alt(4) is an intersecting set of size 12, whereas the point stabilizer in this action has only cardinality 6.

[3] For instance, in the projective general linear group $G = \mathrm{PGL}_d(q)$ in its 2-transitive action on the $(q^d - 1)/(q - 1)$ points of the projective space $\mathrm{PG}_{d-1}(q)$, the intersecting sets of maximal cardinality are either cosets of the stabilizer of a point or cosets of the stabilizer of an hyperplane, see [88].

[4] Observe that all point stabilizers have the same cardinality when G is transitive.

The *intersection density* of G is

$$\rho(G) = \max\{\rho(\mathcal{F}) \mid \mathcal{F} \subseteq G, \mathcal{F} \text{ is intersecting}\}.$$

This invariant was introduced by Li, Song and Pantagi in [70] to measure how "close" G is from satisfying the Erdős–Ko–Rado theorem.

Let $\mathcal{D}$ be the set of all *derangements* of G, where a derangement is a permutation without fixed points. The *derangement graph* of G is the graph Γ_G whose vertex set is the set G and whose edge set consists of all pairs $(h, g) \in G \times G$ such that $gh^{-1} \in \mathcal{D}$. Thus, Γ_G is the Cayley graph of G with connection set $\mathcal{D}$. With this terminology, an intersecting family of G is an *independent set* or *coclique* of Γ_G, and vice versa. As customary, we denote by $\omega(\Gamma_G)$ the maximal size of a clique and by $\alpha(\Gamma_G)$ the maximal size of a coclique.

Now, the clique-coclique bound [48, Theorem 2.1.1]

$$\alpha(\Gamma_G)\omega(\Gamma_G) \leq |V\Gamma_G| = |G| \tag{1.7}$$

can be used to extract useful information on the intersection density of G. Indeed, from (1.7) and from the definition of intersection density, we obtain

$$\rho(G) \leq \frac{|\Omega|}{\omega(\Gamma_G)}. \tag{1.8}$$

When G is transitive and $|\Omega| \geq 2$, Jordan's theorem ensures that G has a derangement g and hence $\{1, g\}$ is a clique of Γ_G of cardinality 2. Therefore, (1.8) yields $\rho(G) \leq |\Omega|/2$.

Theorem 1.5 in [76] shows that, when G is transitive and $|\Omega| \geq 3$, the derangement graph Γ_G has a triangle and hence $\rho(G) \leq |\Omega|/3$. Despite the simplicity of the proof of Jordan's theorem, the proof of this result is quite involved and ultimately relies on the Classification of the Finite Simple Groups and on considerations on weak normal coverings.

In the light of these two results, Question 6.1 in [76] asks for the existence of a function $f : \mathbb{N} \to \mathbb{N}$ such that, if G is transitive of degree n and Γ_G has no clique of cardinality k, then $n \leq f(k)$. Indeed, when $k = 2$, we have $n \leq 1$ (from Jordan's theorem) and, when $k = 3$, we have $n \leq 2$ (from [76, Theorem 1.5]).

Using some of the results in this monograph, we show that the derangement graph Γ_G of a transitive group of degree at least 3 cannot be bipartite; this is weaker than proving that Γ_G has a triangle, but nevertheless it gives an idea of how normal weak coverings come into play.

Proposition 1.14 *Let G be a transitive group of degree $n \geq 3$. Then the derangement graph Γ_G is not bipartite.*

Sketch of the Proof We argue by contradiction and we let G be a transitive permutation group on Ω with $|\Omega| \geq 3$ such that the derangement graph Γ_G of

1.4 The Derangement Graph

G is bipartite.[5] Among all counterexamples, we choose G so that $|\Omega|$ is as small as possible.

We fix a bipartition

$$\mathcal{B} = \{H, G \setminus H\}$$

of the vertices of Γ_G. Without loss of generality suppose that H contains the identity element of G. Since Γ_G is bipartite, this implies that no elements of H are derangements. The group G acts as a group of automorphisms on the graph Γ_G via its right regular representation. Since G acts transitively on the vertices of Γ_G, the subgroup $G_\mathcal{B}$ of G fixing setwise the two parts of the bipartition H and $G \setminus H$ has index 2 in G and acts transitively on both H and $G \setminus H$. As $1 \in H$, we deduce

$$H = \{1^x \mid x \in G_\mathcal{B}\} = \{1 \cdot x \mid x \in G_\mathcal{B}\} = G_\mathcal{B}.$$

This shows that H is a subgroup of G with

$$[G : H] = 2 \text{ and } H \trianglelefteq G. \tag{1.9}$$

As usual, let $\mathcal{D}$ be the set of derangements of G. The subgroup $\langle \mathcal{D} \rangle$ of G generated by $\mathcal{D}$ contains $G \setminus H$ and also the identity element of G. Therefore, $|\langle \mathcal{D} \rangle| \geq |G|/2 + 1$. This shows that

$$G = \langle \mathcal{D} \rangle. \tag{1.10}$$

From (1.10), we deduce that the Cayley graph Γ_G is connected. Since $H \trianglelefteq G$ by (1.9), $G_\omega H$ is a subgroup of G, for every $\omega \in \Omega$. As H has no derangements, by the theorem of Jordan, H cannot be transitive on Ω. Thus H is intransitive on Ω and $G_\omega H$ is a proper subgroup of G. However, $H \leq G_\omega H < G$ and $[G : H] = 2$; therefore $H = HG_\omega$ and $G_\omega \leq H$. This yields

$$G_\omega \leq H, \quad \forall \omega \in \Omega. \tag{1.11}$$

In particular, $G_\omega = H_\omega$, for all $\omega \in \Omega$. From (1.11) and from the fact that H is intersecting, we deduce

$$H = \bigcup_{\omega \in \Omega} G_\omega = \bigcup_{\omega \in \Omega} H_\omega. \tag{1.12}$$

---

[5] We do not give the entire proof of this result, because in part it relies on the O'Nan–Scott theorem classifying finite primitive permutation groups. Hence, giving the whole proof of this proposition would take us too far astray. However, we give enough information to understand the relevance of (weak) normal coverings in the study of density of derangement graphs.

For the rest of the proof we fix $\omega \in \Omega$ and $g' \in G \setminus H$ and we set $\omega' := \omega^g$. As $[G : H] = 2$ and as H is intransitive on Ω, we deduce that H has two orbits on Ω; namely

$$\Delta := \omega^H = \{\omega^h \mid h \in H\} \text{ and } \Delta' := \omega'^H = \{\omega'^h \mid h \in H\}. \quad (1.13)$$

From (1.12) and (1.13), we deduce

$$H = \bigcup_{h \in H} H_\omega^h \cup \bigcup_{h \in H} H_{\omega'}^h.$$

In other words, H has two subgroups H_ω and $H_{\omega'}$ such that the H-conjugates of H_ω and $H_{\omega'}$ cover the whole of H.

Suppose now that $H_\omega = H$ (or that $H_{\omega'} = H$). As $H_\omega = G_\omega$, we deduce $G_\omega = H$. Since G is a transitive permutation group on Ω, G_ω is a core-free subgroup of G, that is, the only normal subgroup of G contained in G_ω is the identity subgroup. However, from (1.9), we have $G_\omega = H \trianglelefteq G$ and hence $H = G_\omega = 1$. This gives $|G| = [G : H] = 2$ and hence $|\Omega| = 2$, which is a contradiction. In particular,

H_ω and $H_{\omega'}$ are proper subgroups of H.

Summing up, we have shown that $\gamma(H) = 2$; moreover, as H_ω and $H_{\omega'}$ are G-conjugate, we have $\gamma_w(H) = 1$.

It can be shown that the minimality of $|\Omega|$ implies that H acts primitively on both Δ and Δ'. At this point, the proof requires the O'Nan–Scott theorem classifying the finite primitive permutation groups. The case of greatest interest is when H is almost simple and, in this case, we may immediately apply Theorem 1.9 with $H = X$ to obtain a final contradiction.[6] □

1.5 Further on Derangements: The Boston-Shalev Conjecture

In addition to exploring Erdős–Ko–Rado-type theorems and derangement graphs, this monograph also establishes connections with the Boston-Shalev conjecture, a topic of considerable interest among researchers in finite permutation groups.

The conjecture posits that, for any finite simple group G acting faithfully and transitively on a set Ω, the proportion of derangements is bounded away from zero by a constant $\delta > 0$, that is,

$$\frac{|\{g \in G \mid g \text{ derangement on } \Omega\}|}{|G|} > \delta.$$

---

[6] Case (2) in Theorem 1.9 does not arise because H_ω does not contain the socle of H.

This hypothesis was initially proposed independently by Boston et al. [7] and Shalev [87], prompting extensive investigations into its validity.

Through a remarkable effort [37–39], Fulman and Guralnick ultimately confirmed the Boston-Shalev conjecture in [40]. This achievement holds some implications for the present work. Specifically, consider a non-abelian simple group G acting transitively on a set Ω, with $\omega_0 \in \Omega$. The elements in $G \setminus \bigcup_{g \in G} G_{\omega_0}^g$ constitute the derangements of G in its action on Ω. The conjecture's validity implies that this set encompasses a substantial proportion of the elements in G. Consequently, it seems conceivable, albeit heuristically, that only in certain very special scenarios, there exists a proper subgroup of G whose conjugates collectively cover all these derangements, thereby exhibiting a normal 2-covering.

The precision of this heuristic speculation is substantially clarified by Theorem 1.3. By systematically classifying all normal 2-coverings of non-abelian simple groups, the theorem furnishes a comprehensive description of the primitive actions of such groups.

1.6 Normal 2-Coverings for Arbitrary Finite Groups

Garonzi and Lucchini [42] have investigated finite groups G with $\gamma(G) = 2$. If $\gamma(G/N) = 2$, for some non-trivial normal subgroup N of G, then we may inductively obtain some information on G and on its normal 2-coverings from the smaller group G/N. Therefore, the most interesting case to consider is when $\gamma(G) = 2$, but $\gamma(G/N) > 2$ for every non-trivial normal subgroup N of G. This is exactly the case studied in [42].

Theorem 1.15 ([42, Theorem 5]) *Let G be a finite group with $\gamma(G) = 2$, but $\gamma(G/N) > 2$ for every non-trivial normal subgroup N of G. Then G has a unique minimal normal subgroup. Moreover, if G is covered with the conjugates of two maximal subgroups, then either one of these two subgroups contains the socle of G or G is an almost simple group.*

Let G be as in the statement of Theorem 1.15 and let N be the unique minimal normal subgroup of G. Since $\gamma(G) = 2$, G has a normal covering $\{H, K\}$ with H and K maximal subgroups of G. From Theorem 1.15, we have that either

- G is almost simple and $N \not\leq H$ and $N \not\leq K$, or
- $N \leq H$ or $N \leq K$.

In the first case, Theorem 1.3 gives a complete description of G because $\{N \cap H, N \cap K\}$ is a weak normal 2-covering of N, see Sect. 1.2 for more details. The second case is more intriguing. Indeed, already in the case that G is almost simple, we do not have a definite answer, see Theorem 12.5.

Interesting examples of the second case, with G not almost simple are in [42, Section 3]. We describe these examples depending on whether N is abelian or not.

In either case, as N is a minimal normal subgroup of G, N is isomorphic to the direct product of pair-wise isomorphic simple groups.

Before describing the examples when N is abelian, we need some notation and a definition. Let M be an elementary abelian group, let K be an irreducible subgroup of $\mathrm{Aut}(M)$ and let

$$K^* = \{k \in K \mid \mathbf{C}_M(k) \neq 1\}.$$

The group K is said to be *almost transitive* if there exists a proper subgroup T of K with

$$K^* \subseteq \bigcup_{x \in K} T^x.$$

Observe that, in this situation, $\gamma(M \rtimes K) = 2$ and $\{K, M \rtimes T\}$ is a normal 2-cover of $M \rtimes K$. Indeed, let $mk \in M \rtimes K$. If $k \in \bigcup_{x \in K} T^x$, then mk is conjugate to an element of $M \rtimes T$. If $k \notin \bigcup_{x \in K} T^x$, then $k \notin K^*$ and hence $\mathbf{C}_M(k) = 1$. Thus $kM = \{k^m \mid m \in M\}$ and hence mk is conjugate to an element of K. Actually, the converse also holds.

Theorem 1.16 ([42, Corollary 14]) *Let G be a finite group with $\gamma(G) = 2$ and with $\gamma(G/X) > 2$ for every non-identity normal subgroup X of G. Let N be the socle of G and assume N is abelian. Then $G = N \rtimes K$, where K is an almost transitive irreducible subgroup of $\mathrm{Aut}(N)$.*

As already observed in [42]:

> In virtue of the previous result, it should be interesting to classify the almost transitive irreducible groups.

When K acts fixed-point-freely on N or when K acts transitively by conjugation on $N \setminus \{1\}$, K is almost transitive. More, exotic examples of almost transitive irreducible groups are in [42, Section 3].

We now describe some examples with N non-abelian. Let S be a finite non-abelian simple group, let p be a prime with $\gcd(p, |S|) = 1$ and let $G = S \mathrm{wr} \langle \sigma \rangle$ with $\sigma = (1\,2\,\ldots\,p) \in \mathrm{Sym}(p)$. Let $N = S^p$ be the base of the wreath product and let

$$H = \{(s, \ldots, s)\sigma^i \mid s \in S, i \in \{0, \ldots, p-1\}\}$$

be a maximal subgroup of G of diagonal type. We show that N and H are the components of a normal 2-covering of G. Indeed, let $g = (s_1, \ldots, s_p)\sigma^i \in G$. When $\sigma^i = 1$, we have $g \in N$. When $\sigma^i \neq 1$, as p is prime, raising g to a suitable power, we may suppose that $i = 1$. Since p is relatively prime to $|S|$, there exists $s \in S$ with

$$s^p = s_1 s_2 \cdots s_p.$$

1.6 Normal 2-Coverings for Arbitrary Finite Groups

Now, let

$$x_1 = 1,\ x_2 = s^{-1}s_1,\ x_3 = s^{-2}s_1s_2,\ \ldots,\ x_p = s^{-(p-1)}s_1s_2\cdots s_{p-1}.$$

We have

$$\begin{aligned}((s,s,\ldots,s)\sigma)^{(x_1,\ldots,x_p)} &= (x_1^{-1}sx_2, x_2^{-1}sx_3, \ldots, x_p^{-1}sx_1)\sigma \\ &= (s_1, s_2, \ldots, s_p)\sigma = g\end{aligned}$$

and hence g is conjugate to an element of H. These are not the only examples arising with N non-abelian. For instance, in [42, Section 3], there are examples of groups where the maximal subgroup H is of product type.

As we already observed in Sect. 1.2, there are also examples of almost simple type. For instance, the group $A := \mathrm{P\Gamma L}_2(8)$ admits a normal 2-covering using its socle $\mathrm{PSL}_2(8)$ and $\mathbf{N}_A(P)$, where P is a 3-Sylow of A. See Theorem 12.5 for more details.

Garonzi and Lucchini [42] explored finite groups G possessing a normal 2-covering, where no proper quotient of G exhibits such a covering. Their investigation offered a comprehensive overview of these groups, delineating that such groups fall into distinct categories: almost simple, affine, product action, or diagonal. Besides the work in this monograph for almost simple groups, at the moment [41] is the only paper dealing with a classification of the remaining categories. Indeed, the authors of [41] present a thorough classification of finite diagonal groups possessing a normal 2-covering, with the attribute that no proper quotient of G has such a covering.

To give details of this classification we first need an example.

Example 1.17 Let T be a non-abelian simple group, let U be a subgroup of $\mathrm{Aut}(T)$ containing the inner automorphisms of T and let p be a prime number with $\gcd(|U|, p) = 1$. Here we identify T with the inner automorphism group of T. Let $N = T^p$, let

$$H = \{(x_1, \ldots, x_p) \in U^p \mid x_i \equiv x_j \pmod{T}, \forall i, j \in \{1, \ldots, p\}\}$$

and let $\sigma = (1\ 2\ \cdots\ p)$ be the cyclic permutation of degree p.

We define

$$G = H \rtimes \langle \sigma \rangle,$$

where σ acts as a group of automorphisms on H by setting

$$(x_1, x_2, \ldots, x_p)^\sigma = (x_2, x_3, \ldots, x_p, x_1),$$

for each $(x_1, \ldots, x_\ell) \in H$.

Finally we let

$$K = \{(x, \ldots, x) \in H \mid x \in U\} \times \langle \sigma \rangle \leq G.$$

Observe that, as $\gcd(|U|, p) = 1$, a Sylow p-subgroup of G has order p and hence $\langle \sigma \rangle$ is a Sylow p-subgroup of G. Moreover, $\mathbf{C}_G(\sigma) = K$.

We claim that H and K are the components of a normal 2-covering of G. Indeed, let $g \in G$. Therefore, $g = (x_1, \ldots, x_p)\sigma^i$, for some $(x_1, \ldots, x_p) \in H$ and $i \in \{0, \ldots, p-1\}$. If $i = 0$, then $g \in H$. Assume then $i \neq 0$. As $i \neq 0$, g has order divisible by p and hence g centralizes a Sylow p-subgroup P of G. By Sylow's theorem, there exists $z \in G$ with $P^z = \langle \sigma \rangle$. Thus g^z centralizes $P^z = \langle \sigma \rangle$ and hence $g^z \in \mathbf{C}_G(\sigma) = K$.

We are now ready to describe the diagonal groups having normal covering number 2.

Theorem 1.18 *Let G be a group with $\gamma(G) = 2$ and $\gamma(G/N) > 2$, for every non-identity normal subgroup N of G. Let H and K be maximal subgroups of G witnessing that $\gamma(G) = 2$. Then either H or K contains the socle of G.*

Let H be the component containing the socle of G. If G is of diagonal type, then G, K and H are isomorphic to one of the groups described in Example 1.17.

1.7 Normal Coverings and Kronecker Classes

There are some remarkable connections between normal coverings and algebraic number fields, see for instance [61, 67, 68, 79].

Given an algebraic number field k and a finite extension field K of k the **Kronecker set** of K over k is defined as the set of all prime ideals of the ring of integers of k having a prime divisor of relative degree one in K. Then, two finite extensions of k are said to be **Kronecker equivalent** if their Kronecker sets have finite symmetric difference, that is, the Kronecker sets differ only in at most a finite number of primes. This defines an equivalence relation and such extensions are said to belong to the same **Kronecker class**. Clearly, extensions in the same Kronecker class have strong arithmetical similarities.

The connection between problems about Kronecker classes in field extensions and group theoretic problems is explained in [61, 67, 79]. Let K and K' be finite extensions of a given fixed algebraic number field k and let M be a Galois extension of k containing K and K'. Let $G = \mathrm{Gal}(M/k)$, $U = \mathrm{Gal}(G/K)$ and $U' = \mathrm{Gal}(M/K')$, in particular, U and U' are the subgroups of G corresponding to K and K' via the Galois correspondence. It is shown in [61, 67] that K and K' are Kronecker equivalent if and only if

$$\bigcup_{g \in G} U^g = \bigcup_{g \in G} U'^g. \tag{1.14}$$

1.7 Normal Coverings and Kronecker Classes

This already gives a very strong connection between the problem of understanding Kronecker classes and natural questions in finite permutation groups. For instance, if we consider the permutation representations of G on the right cosets of U and on the right cosets of U', then (1.14) is equivalent to the fact that in these two permutation representations of G the set of derangements is the same.

There is one special case where (1.14) yields a natural connection with (weak) normal coverings. Indeed, the special case where K'/k is a Galois extension and K is an extension of K' corresponds to $U \leq U' \trianglelefteq G$. In particular, in this special case, K/k and K'/k are Kronecker equivalent if and only if

$$U' = \bigcup_{g \in G} U^g.$$

Using the terminology in [79], this yields that U' is a G-covering of U. As G acts by conjugation as a group of automorphisms on U', when $U' \neq U$, we deduce that $\{U\}$ is a weak normal 1-covering of U'.

There is a number of problems arising in Kronecker classes in algebraic number fields that have been addressed using finite group theory. We report here some open conjectures.

Conjecture 1.19 (Neumann, Praeger, See [82]) There is an integer function f such that, if G is a finite group with subgroups U, U' such that $|G : U'| = n$ and

$$\bigcup_{g \in G} U^g = \bigcup_{g \in G} U'^g,$$

then $|G : U| \leq f(n)$.

This conjecture phrased in terms of Kronecker classes is as follows.

Conjecture 1.20 There is an integer function f such that, if K/k is an extension of degree n of algebraic number fields and L/k is Kronecker equivalent to K/k, then $|L : k| \leq f(n)$.

As we mentioned above, with respect to (weak) normal coverings, the case of particular interest is when $U \leq U' \trianglelefteq G$.

Conjecture 1.21 (Neumann, Praeger, See [82]) There exists an integer function g such that, if U' is a finite group, G is a group of automorphisms of U' containing $\text{Inn}(U')$ the inner automorphisms of U' and with $n = |G : \text{Inn}(U')|$, and U is a subgroup of U' with

$$U' = \bigcup_{g \in G} U^g,$$

then $|U' : U| \leq g(n)$.

1.8 Structure of the Book and Comments

In our proof of Theorems 1.3 and 1.5, we use the Classification of the Finite Simple Groups. Thus in Table 1.3 we have collected the alternating groups, the exceptional groups of Lie type and the sporadic simple groups, in Tables 1.4, 1.5, 1.6, and 1.7 we have collected the classical groups. We take into account the various isomorphisms among finite simple groups. For instance, we do not list the weak normal 2-coverings of $PSL_2(4)$, $PSL_2(5)$, $PSL_2(9)$ and $PSp_4(2)'$, because these already appear among the alternating groups.

The classification of the alternating groups admitting normal covering number 2 appears in [13]. There it is shown that the alternating group A_n has normal covering number 2 if and only if $4 \leq n \leq 8$. In this monograph, we do not repeat the arguments in [13] to prove our more general result for weak normal coverings. Indeed, it is easily checked that the whole reasoning in [13] works for our more general classification apart for the case A_9, which is one of the groups appearing in Corollary 1.6. The heart of the matter is that the 9-cycles in A_9 split into two conjugacy classes while the 9-cycles in $P\Gamma L_2(8)$ are all S_9-conjugate. Thus

Table 1.3 Weak normal 2-coverings of non-abelian simple groups: alternating, exceptional and sporadic simple groups

Grp	Comp. H	Comp. K	Comments	N.	Nr.
A_5	$A_5 \cap (S_2 \times S_3)$	D_{10}		1	2
	A_4	D_{10}		1	
A_6	$A_6 \cap (S_2 \times S_4)$	A_5		2	2
	$A_6 \cap (S_3 \mathrm{wr} S_2)$	A_5		2	
A_7	$A_7 \cap (S_2 \times S_5)$	$SL_3(2)$		2	1
A_8	$A_8 \cap (S_3 \times S_5)$	$2^3 : SL_3(2)$	$A_8 \cong PSL_4(2)$	2	1
A_9	$A_9 \cap (S_4 \times S_5)$	$P\Gamma L_2(8)$		0	1
$G_2(q)$	$SL_3(q).2$	$SU_3(q).2$	$q \geq 4$, q even	1	1
$G_2(2)'$	$PSL_2(7)$	$4 \cdot S_4$	$G_2(2)' \cong PSU_3(3)$	1	2
	$PSL_2(7)$	$3_+^{1+2} : 8$		1	
$G_2(3)$	$PSL_2(13)$	$[q^5] : GL_2(3)$		0	2
	$PSL_3(3) : 2$	$PSL_2(8) : 3$		0	
$^2G_2(3)'$	D_{18}	D_{14}	$^2G_2(3)' \cong PSL_2(8)$	1	2
	D_{18}	$2^3 : 7$		1	
$^2F_4(2)'$	$2.[2^8] : 5 : 4$	$PSL_3(3) : 2$		1	2
	$2.[2^8] : 5 : 4$	$PSL_2(25)$		1	
$F_4(q)$	$^3D_4(q).3$	$Spin_9(q)$	$q = 3^a$	1	1
M_{11}	$M_8 : S_3 \cong 2 \cdot S_4$	$PSL_2(11)$	Notation from [29]	1	3
	$M_9 : S_2 \cong 3^2 : Q_8.2$	$PSL_2(11)$		1	
	$M_{10} \cong A_6.2$	$PSL_2(11)$		1	
M_{12}	$M_{10} : 2 \cong A_6.2^2$	$PSL_2(11)$	Notation from [29]	0	2
	M_{11}	$2 \times S_5$		0	

1.8 Structure of the Book and Comments

Table 1.4 Weak normal 2-coverings of non-abelian simple groups: linear groups (For $PSL_2(4)$, $PSL_2(5)$, $PSL_2(9)$, see Table 1.3)

Grp	Comp. H	Comp. K	Comments	N.	Nr.
$PSL_2(7)$	S_4	Borel		2	1
$PSL_2(q)$	D_{q+1}	Borel	$q > 9$, q odd	1	1
$PSL_2(q)$	$D_{2(q+1)}$	Borel	$q > 4$, q even	1	2
	$D_{2(q+1)}$	$D_{2(q-1)}$	$q > 4$, q even	1	
$PSL_3(q)$	$\left(\frac{q^2+q+1}{\gcd(3,q-1)}\right):3$	Max. parabolic	$q \neq 4$	2	1
$PSL_3(4)$	$SL_3(2)$	A_6		0	2
	$SL_3(2)$	Max. parabolic		6	
$PSL_4(q)$	$\frac{1}{d}SL_2(q^2).(q+1).2$	$\frac{1}{d}E_q^3 : GL_3(q)$	$d := \gcd(4, q-1)$	2	1

Table 1.5 Weak normal 2-coverings of non-abelian simple groups: unitary groups

Grp	Comp. H	Comp. K	Comments	N.	Nr.
$PSU_3(q)$	$(q^2 - q + 1):3$	$GU_2(q)$	$q = 3^a, a > 1$	1	1
$PSU_3(3)$	$PSL_2(7)$	$GU_2(3)$		1	2
	$PSL_2(7)$	$E_3^{1+2}:8$		1	
$PSU_3(5)$	A_7	$\frac{1}{3}GU_2(5)$		0	2
	A_7	$\frac{1}{3}E_5^{1+2}:24$		3	
$PSU_4(q)$	$\frac{1}{d}GU_3(q)$	$\frac{1}{d}E_q^4 : SL_2(q^2) : (q-1)$	$d := \gcd(4, q+1)$	1	1
			$q \geq 4$		
$PSU_4(2)$	$GU_3(2)$	$Sp_4(2)$		1	2
	$GU_3(2)$	$E_2^4 : SL_2(4)$		1	
$PSU_4(3)$	A_7	$\frac{1}{4}E_3^{1+4} : SU_2(3) : 8$		4	4
	$\frac{1}{4}GU_3(3)$	$\frac{1}{4}E_3^4 : SL_2(9):2$		1	
	$PSL_3(4)$	$\frac{1}{4}E_3^{1+4} : SU_2(3) : 8$		2	
	$\frac{1}{4}GU_3(3)$	$PSU_4(2)$		0	
$PSU_6(2)$	$Sp_6(2)$	$PGU_5(2)$		0	2
	$PSU_4(3)$	$PGU_5(2)$		0	

$\bigcup_{g \in A_9} P\Gamma L_2(8)^g$ does not contain all the 9-cycles of A_9, while $\bigcup_{g \in S_9} P\Gamma L_2(8)^g$ does (for details see [13, pp. 17–18]).

The classification of the sporadic simple groups having normal covering number 2 is in [77] and it is carried out in two parts. In the first general part, the arguments are via group orders and this part applies verbatim to our situation. In the second part the author analyses, often with the use of a computer, the exceptional cases not covered by the previous investigation. For these exceptional cases, reported in [77, Table 2] we have used a computer for finding weak normal 2-coverings. The only exceptional case that appears is the Mathieu group M_{12}.

In [16, 73, 77], the authors classify the finite simple exceptional groups of Lie type having covering number 2. Again, we do not replicate the arguments in these

Table 1.6 Weak normal 2-coverings of non-abelian simple groups: symplectic groups (For $PSp_4(2)' \cong A_6$, see Table 1.3)

Grp	Comp. H	Comp. K	Comments	N.	Nr.
$PSp_4(3)$	$\frac{1}{2}E_3^{1+2} : (2 \times Sp_2(3))$	$PSp_2(9) : 2$		1	2
	$\frac{1}{2}E_3^{1+2} : (2 \times Sp_2(3))$	$2^4.A_5$		1	
$Sp_n(q)$	$SO_n^-(q)$	$SO_n^+(q)$	$n \geq 6$	1	1
			q even		
$Sp_4(q)$	$SO_4^-(q)$	$SO_4^+(q)$	q even	2	1
			$q > 4$		
$Sp_4(4)$	$SO_4^-(4)$	$SO_4^+(4)$		2	2
	$Sp_2(16) : 2$	$Sp_4(2)$		0	
$PSp_6(3^f)$	$\frac{1}{2}(Sp_2(3^f) \perp Sp_4(3^f))$	$\frac{1}{2}Sp_2(3^{3f}) : 3$		1	1

Table 1.7 Weak normal 2-coverings of non-abelian simple groups: orthogonal groups. Observe that the structure of the maximal parabolic subgroups appearing as components for $P\Omega_8^+(3)$ is described in the "Comments" column

Grp	Comp. H	Comp. K	Comments	N.	Nr.
$P\Omega_8^+(2)$	A_9	$E_2^{1+8} : (GL_2(2) \times \Omega_4^+(2))$	K stb. t.s. ln	0	4
	$Sp_6(2)$	$E_2^{1+8} : (GL_2(2) \times \Omega_4^+(2))$	K stb. t.s. ln	0	
	$Sp_6(2)$	$E_2^6 : \Omega_6^+(2)$	K stb. t.s pnt	0	
	$E_2^6 : \Omega_6^+(2)$	$(\Omega_2^-(2) \times \Omega_6^-(2)).2$	H stb. t.s. pnt	0	
$P\Omega_8^+(3)$	$\Omega_7(3)$	$\frac{1}{2}(\Omega_3(3) \times \Omega_5(3)).[4]$		0	4
	$\Omega_7(3)$	Max. parabolic	K stb. t.s. ln	0	
	$\Omega_7(3)$	Max. parabolic	K stb. t.s. pnt	0	
	Max. parabolic	$\frac{1}{2}(\Omega_2^-(3) \times \Omega_6^-(3)).[4]$	H stb. t.s. pnt	0	

papers. We simply observe that, with minor modifications, the reasoning in [73, 77] gives a proof to Theorem 1.3 in the case of finite simple exceptional groups of Lie type. The only exceptional case arising here is $G_2(3)$. The study of $G_2(3)$ in [77] was carried via a computer computation and ours too.

Therefore, the bulk of our work is the proof of Theorems 1.3 and 1.5 for simple classical groups. In [14], it has been proved that, if the projective special linear group $PSL_n(q)$ admits a normal 2-covering then $2 \leq n \leq 4$. The same proof yields the same result for weak normal 2-coverings. Since in [14] it is also shown that for $2 \leq n \leq 4$ there exists a normal 2-covering of $PSL_n(q)$, we deduce that $\gamma_w(PSL_n(q)) \leq 2$ for those n. Thus, in dealing with the simple groups $PSL_n(q)$, we may suppose that $2 \leq n \leq 4$ and concentrate on the explicit description of all the possible maximal components of a weak normal 2-covering and on excluding $\gamma_w(PSL_n(q)) = 1$.

The structure of the book is straightforward. In Chap. 2, we give some basic preliminaries which will also serve useful for setting some notation. Then, in Chap. 3, we consider the linear groups $PSL_n(q)$; in Chap. 4, we consider the unitary groups $PSU_n(q)$; in Chap. 5, we consider the symplectic groups $PSp_n(q)$;

1.8 Structure of the Book and Comments

in Chap. 6, we consider the odd dimensional orthogonal groups $P\Omega_n(q)$; in Chaps. 7 and 8, we deal with even dimensional orthogonal groups $P\Omega_n^-(q)$ and $P\Omega_n^+(q)$; in Chap. 9, we prove Theorem 1.9; in Chap. 10, we investigate the almost simple groups with sporadic socle giving the details for obtaining Table 1.2.

In Chap. 11, we use Theorem 1.5 to classify the weak normal 2-coverings and the normal 2-coverings of the non-abelian simple groups. This means that we drop the hypothesis about the maximality of the components and give the description of all components in a weak normal 2-covering of G, when G is a non-abelian simple group with $\gamma_w(G) = 2$. Our main result is Theorem 11.1 and to our surprise most weak normal 2-coverings do require at least one of the components to be maximal, see Corollary 11.2.

In Chap. 12, we study the degenerate normal 2-coverings of almost simple groups.

Chapter 2
Preliminaries

Given a group G, we denote as customary its derived group by G'. A group is called *perfect* when $G' = G$.

In this chapter, we list a number of remarks and results that are needed throughout the whole monograph. These facts are divided in various sections depending on the particular aspect we are focusing on. For instance, we give some details on the groups we are dealing with and their actions on the underlying vector spaces. We will also need some arithmetical observations.

2.1 Classical Groups

Throughout the book let n, f be positive integers, p be a prime and $q = p^f$. We consider the following *classical groups* $\tilde{G}$ of dimension n:

- $\mathrm{SL}_n(q)$ with $n \geq 1$,
- $\mathrm{SU}_n(q)$ with $n \geq 1$,
- $\mathrm{Sp}_n(q)$ with n even and $n \geq 2$,
- $\Omega_n(q)$ with qn odd and $n \geq 1$, and
- $\Omega_n^\pm(q)$ with n even and $n \geq 2$.

The corresponding *simple classical groups* $G := \tilde{G}/Z(\tilde{G})$ are

$$\mathrm{PSL}_n(q),\ \mathrm{PSU}_n(q),\ \mathrm{PSp}_n(q),\ \mathrm{P}\Omega_n(q),\ \text{and}\ \mathrm{P}\Omega_n^\pm(q),$$

where we consider only $n \geq 2$ for linear groups, $n \geq 3$ for unitary groups, $n \geq 4$ for symplectic groups, $n \geq 7$ for odd dimensional orthogonal groups and $n \geq 8$ for even dimensional orthogonal groups. With the restrictions on n as above, these are indeed non-abelian simple groups, except for $\mathrm{PSL}_2(2)$, $\mathrm{PSL}_2(3)$, $\mathrm{PSU}_3(2)$ and

PSp$_4$(2). Moreover, taking into account the various isomorphisms among simple groups (see [66, Section 2.9]), the choices for n guarantee that every finite non-abelian simple group is considered just one time.

For some of our preliminary results or some arguments, we need to deal with arbitrary classical groups as defined above and hence with no restrictions on n. However, recall that our main results are only concerned with non-abelian simple groups.

We also need the ***general classical groups***, which we denote by $\hat{G}$, denoted by

$$\text{GL}_n(q), \ \text{GU}_n(q), \ \text{Sp}_n(q), \ \text{O}_n(q), \ \text{and} \ \text{O}_n^\pm(q).$$

For the groups $\hat{G}$ we adopt the same limitations for n adopted for the corresponding $\tilde{G}$. As usual, for the notation we follow [66, Section 2.1]. Recall that $\hat{G}' = \tilde{G}$.

2.2 Normal and Weak Normal Coverings of Classical and Simple Classical Groups

Let $\tilde{G}$ be a classical group such that G is a non-abelian simple group. Then $\tilde{G}$ is a perfect group, see for instance [66, Proposition 2.9.2 (ii)]. As a consequence if H is a proper subgroup of $\tilde{G}$, then $HZ(\tilde{G})$ is also a proper subgroup of $\tilde{G}$. In particular, every maximal subgroup of $\tilde{G}$ contains $Z(\tilde{G})$. We illustrate the natural link between coverings of $\tilde{G}$ and coverings of G. That link is extensively used throughout the monograph. To begin with, if $\pi : \tilde{G} \to G$ is the natural projection of $\tilde{G}$ onto G and $X \le G$, we define $\tilde{X} := \pi^{-1}(X)$. Of course, $\tilde{X}$ is a subgroup of $\tilde{G}$ containing $Z(\tilde{G})$. Note also that $\pi(\tilde{X}) = X$ and that X is maximal in G if and only if $\tilde{X}$ is maximal in $\tilde{G}$.

Let k be a positive integer and $\mu = \{H_1, \ldots, H_k\}$ be a set of distinct proper subgroups of G. Then, $\tilde{\mu} := \{\tilde{H}_1, \ldots, \tilde{H}_k\}$ is a set of distinct proper subgroups of $\tilde{G}$ containing $Z(\tilde{G})$ and the map $\mu \mapsto \tilde{\mu}$ establishes a bijection between sets of k distinct proper subgroups of G and sets of k distinct proper subgroups of $\tilde{G}$ containing $Z(\tilde{G})$. It is immediate to check that μ is a normal covering of G if and only if $\tilde{\mu}$ is a normal covering of $\tilde{G}$ with components containing $Z(\tilde{G})$. Hence, if $\tilde{\mu}$ is a normal covering of $\tilde{G}$ with maximal components, then μ is a normal covering of G (with maximal components). It follows that

$$\gamma(\tilde{G}) = \gamma(G).$$

Moreover our main result can be used to determine the non-solvable classical groups having covering number 2 and to characterize the maximal components of their normal 2-coverings.

We now explore the link between the weak normal coverings of $\tilde{G}$ and G. Since $Z(\tilde{G})$ is a characteristic subgroup of $\tilde{G}$, we have a natural homomorphism of Aut($\tilde{G}$)

in Aut(G). It follows that if $\tilde{\mu}$ is a weak normal covering of $\tilde{G}$ with components containing $Z(\tilde{G})$, then μ is a weak normal covering of G. In particular, we have

$$\gamma_w(\tilde{G}) \geq \gamma_w(G).$$

From [66, Chapter 2], we see that, when G is simple, the homomorphism of Aut($\tilde{G}$) in Aut(G) is surjective, with the only exception of $G = \text{P}\Omega_8^+(q)$ and q odd. Using that fact it easily follows that, apart from that exceptional case, if μ is a weak normal covering of G, then $\tilde{\mu}$ is a weak normal covering of $\tilde{G}$. In particular, apart from that exceptional case, we have

$$\gamma_w(\tilde{G}) = \gamma_w(G) \qquad (2.1)$$

and our main result can be used to determine the non-solvable classical groups having weak normal covering number 2 and to characterize the maximal components of their weak normal 2-coverings.

In other words, apart from the exceptional case, we may freely work between weak normal coverings of G and $\tilde{G}$. Extra care has to be taken when $G = \text{P}\Omega_8^+(q)$ with q odd, because the triality automorphism of G does not lift to an automorphism of $\tilde{G}$. From Table 1.7, we deduce that, among simple groups G having weak normal covering number 2, the only possible exception to (2.1) is for $G := \text{P}\Omega_8^+(3)$. As we all know "the devil is in the details" and in fact, it can be checked with a computer that $\gamma_w(\Omega_8^+(3)) \geq 3 > \gamma_w(\text{P}\Omega_8^+(3)) = 2$.

Theorem 2.1 *Let $\tilde{G}$ be a classical group such that G is a non-abelian simple group. Then the following hold*

- $\gamma(\tilde{G}) = \gamma(G)$,
- $\gamma_w(\tilde{G}) = \gamma_w(G)$ *if and only if $G \not\cong \text{P}\Omega_8^+(3)$.*

2.3 From Weak Normal Coverings to Normal Coverings

In this section, we describe a situation that we often face in our work.

Let G be a non-abelian simple group and let $\{H, K\}$ be a weak normal 2-covering of G. Observe that from the theorem of Saxl [85], H and K are in distinct Aut(G)-classes. Clearly, $\mathcal{C} := \{\{H^\varphi, K^\psi\} \mid \varphi, \psi \in \text{Aut}(G)\}$ is the Aut(G)-class of weak normal 2-coverings with representative $\{H, K\}$. Now, $\mathcal{C}$ is a union of, C say, G-classes of normal 2-coverings. The case $C = 0$ is special and consists in the case where no normal 2-covering can be extracted from this Aut(G)-class. Suppose then $C \geq 1$, that is, there exist $\varphi, \psi \in \text{Aut}(G)$ such that $\{H^\varphi, K^\psi\}$ is a normal 2-covering. For simplicity, we may suppose that $\{H, K\}$ itself is a normal 2-covering. Here, we give a lower and an upper bound for C.

Let $h := |\mathrm{Aut}(G) : G\mathrm{N}_{\mathrm{Aut}(G)}(H)|$ and $k := |\mathrm{Aut}(G) : G\mathrm{N}_{\mathrm{Aut}(G)}(K)|$. In particular, $\{H^\varphi \mid \varphi \in \mathrm{Aut}(G)\}$ is a union of h distinct G-conjugacy classes and similarly $\{K^\varphi \mid \varphi \in \mathrm{Aut}(G)\}$ is a union of k distinct G-conjugacy classes. It is not hard to verify that

$$\max\{h, k\} \text{ divides } C \text{ and } C \leq hk.$$

In particular, when $h = 1$ (respectively $k = 1$), we have $C = k$ (respectively $C = h$). In most of our arguments, we use this basic observation for extracting information about normal 2-coverings from weak normal 2-coverings.

For classical groups and for their maximal subgroups, the values of h and k are in the literature in the "c" column in the tables in [9, 66].

2.4 Action of $\hat{G}$ on Its Natural Module

Set $\delta := 1$ when $\hat{G} \neq \mathrm{GU}_n(q)$ and set $\delta := 2$ when $\hat{G} = \mathrm{GU}_n(q)$. Let $V = (\mathbb{F}_{q^\delta})^n$ be the natural n-dimensional $\mathbb{F}_{q^\delta}\hat{G}$-module consisting of row vectors $x = (x_1, \ldots, x_n)$ in n components $x_i \in \mathbb{F}_{q^\delta}$ and endowed with the suitable $\hat{G}$-invariant form. This form is non-degenerate, with the only exception of $\hat{G} = \mathrm{GL}_n(q)$. If $x \in \mathrm{GL}_n(q)$ admits the eigenvalue $\lambda \in \mathbb{F}_q$, we denote the corresponding eigenspace by $V_\lambda(x)$. When $\lambda = 1$, we also use $\mathbf{C}_V(x)$ to denote $V_1(x)$, because $V_1(x)$ is indeed the centralizer in V of x in the semidirect product $V \rtimes \mathrm{GL}_n(q)$.

In our proofs, for discussing weak normal coverings of $\tilde{G}$, we use the action of $\tilde{G} \leq \hat{G}$ on the natural module V. Observe, that except when $G = \mathrm{PSL}_n(q)$ with $n \geq 3$, $G = \mathrm{Sp}_4(2^f)$ or $G = \mathrm{P}\Omega_8^+(q)$, the group $\mathrm{Aut}(\tilde{G})$ acts on the natural module as a semilinear group and hence we may also discuss the action of the elements of $\mathrm{Aut}(\tilde{G})$ on V. Extra care has to be taken when $G = \mathrm{PSL}_n(q)$ with $n \geq 3$, $G = \mathrm{Sp}_4(2^f)$ and $G = \mathrm{P}\Omega_8^+(q)$.

Given $g \in \hat{G}$, when V is a completely reducible $\mathbb{F}_{q^\delta}\langle g\rangle$-module, that is, V decomposes into the direct sum of irreducible $\mathbb{F}_{q^\delta}\langle g\rangle$-submodules V_i of dimensions d_i, for $i \in \{1, \ldots, k\}$, we say that **the action of** g **is of type** $d_1 \oplus \cdots \oplus d_k$. Note that g acts irreducibly (that is, V is an irreducible $\mathbb{F}_{q^\delta}\langle g\rangle$-module) if and only if its characteristic polynomial is irreducible.

Since we usually work with semisimple elements, we shall use some facts about the maximal tori of the classical groups, in particular their orders and their action on the natural module V, which can be found in [49, Chapter 2 and Section 4.2]. We also make use of the work of Wall [90, 91], which broadly speaking describes the Jordan forms of the matrices belonging to a given classical group. Incidentally, we denote by $\mathbf{o}(g)$ the order of a group element g. Another good source of detailed information on the conjugacy classes in classical groups is [19].

Since we want to determine whether there exist maximal subgroups H, K of G such that any element of G is $\mathrm{Aut}(G)$-conjugate to an element of H or K, we adopt the systematic description of the maximal subgroups of the classical groups given by Aschbacher in [1]. Actually, since Aschbacher does not consider almost simple groups containing a graph automorphism, we use the extended, and slightly different, notation in [66]. The maximal subgroups of G, or of $\hat{G}$, or of any almost simple group with socle G are divided into nine families. The families C_i, with $i \in \{1,\ldots,8\}$, are defined in terms of the geometric properties of their action on V and the main result of Aschbacher states that any maximal subgroup belongs to $\bigcup_{i=1}^{8} C_i$ or to an additional family S, consisting of groups satisfying certain irreducibility conditions. For notation and structure theorems on these maximal subgroups and other details of our investigation, we refer to the book of Kleidman and Liebeck [66] or to the book of Bray et al. [9] when the rank is small. As we mentioned above, [9, 66] take also into account almost simple groups having socle a simple classical group and containing a graph automorphism.

2.5 Huppert's Theorem and Singer Cycles

In this section the vector space V admits a non-degenerate form or quadratic form of classical type which is preserved by the group G. We frequently make use of a theorem of Huppert [60, Satz 2], which we apply to semisimple elements $s \in G$. Such elements generate a subgroup acting completely reducibly on V, and by Huppert's Theorem, V admits an orthogonal decomposition of the following form:

$$V = V_+ \perp V_- \perp ((V_{1,1} \oplus V'_{1,1}) \perp \cdots \perp (V_{1,m_1} \oplus V'_{1,m_1})) \perp \cdots \qquad (2.2)$$
$$\perp ((V_{r,1} \oplus V'_{r,1}) \perp \cdots \perp (V_{r,m_r} \oplus V'_{r,m_r}))$$
$$\perp (V_{r+1,1} \perp \cdots \perp V_{r+1,m_{r+1}}) \perp \cdots \perp (V_{t',1} \perp \cdots \perp V_{t',m_{t'}}),$$

where V_+ and V_- are the eigenspaces of s for the eigenvalues 1 and -1, of dimensions d_+ and d_-, respectively (note $V_\pm$ is non-degenerate if $d_\pm > 0$ and we set $d_- = 0$ if q is even), and each $V_{i,j}$ is an irreducible $\mathbb{F}_{q^\delta}\langle s \rangle$-submodule. Moreover for $i = r+1, \ldots, t'$, $V_{i,j}$ is non-degenerate of dimension $2d_i/\delta$ and s induces an element $y_{i,j}$ of order dividing $q^{d_i} + 1$ on $V_{i,j}$ (in the unitary case $\delta = 2$ and the dimension d_i is odd).

For $i = 1, \ldots, r$, $V_{i,j}$ and $V'_{i,j}$ are totally isotropic of dimension d_i/δ (here d_i is even if $\delta = 2$), $V_{i,j} \oplus V'_{i,j}$ is non-degenerate, and s induces an element $y_{i,j}$ of order dividing $q^{d_i} - 1$ on $V_{i,j}$ while inducing the adjoint representation $(y_{i,j}^{-1})^T$ on $V'_{i,j}$, where x^T denotes the transpose of the matrix x.

For our claims about the orders of the $y_{i,j}$ and for some standard facts on the structure of the maximal tori of the finite classical groups, we also refer to [25, 62].

Some special semisimple elements help our task of identifying the components of a weak normal 2-covering of a classical group. Among them a main role is played by the **Singer cycles**, that is, those elements in $\hat{G}$ or $\tilde{G}$ acting irreducibly on V and having maximum order. These elements were intensively studied by Huppert in [60] and by Hestenes in [57].

We recall now some main facts about Singer cycles. Other information needed for the monograph is given in Sect. 2.8. As usual, let $\mathbb{F}_{q^{\delta n}}$ be the field with $q^{\delta n}$ elements. Let $a \in \mathbb{F}_{q^n}$ and consider the multiplication $\pi_a : \mathbb{F}_{q^n} \to \mathbb{F}_{q^n}$ defined by $\pi_a(x) = ax$ for all $x \in \mathbb{F}_{q^n}$. The maps π_a with $a \neq 0$ form a group isomorphic to $\mathbb{F}_{q^n}^*$. For every $d \mid n$, the set $V := \mathbb{F}_{q^n}$ can be interpreted as a vector space of dimension n/d over the field $\mathbb{F}_{q^d}$ and the map π_a is an $\mathbb{F}_{q^d}$-linear transformation of V. Thus, once a basis is fixed, π_a induces a matrix belonging to $\mathrm{GL}_{n/d}(q^d)$. If $\langle a \rangle = \mathbb{F}_{q^n}^*$, by [59, II. Satz 7.3], we have that

$$\det{}_{\mathbb{F}_{q^d}}(\pi_a) = a^{\frac{q^n-1}{q^d-1}}. \tag{2.3}$$

Only when necessary, or useful for certain purposes, we keep track of the field $\mathbb{F}_{q^d}$ under consideration in the notation of the determinant of π_a.

It is well-known that the Singer cycles in $\mathrm{GL}_n(q)$ are, up to conjugacy, exactly the matrices representing π_a as an $\mathbb{F}_q$-linear transformation of V, for a a generator of $\mathbb{F}_{q^n}^*$. Hence if $s \in \mathrm{GL}_n(q)$ is any Singer cycle, by (2.3), we get $\det(s) = \det{}_{\mathbb{F}_q}(\pi_a) = a^{\frac{q^n-1}{q-1}}$, for some $a \in \mathbb{F}_{q^n}^*$ such that $\langle a \rangle = \mathbb{F}_{q^n}^*$. Thus $\det(s)$ has order $q-1$ and generates $\mathbb{F}_q^*$.

There is a useful construction of Singer cycles by suitable multiplications also for the other general classical groups.

For the unitary case, with n odd, let $\langle a \rangle = \mathbb{F}_{q^{2n}}^*$. Then, by [57, Theorem 5.2], the multiplication $\pi_{a^{q^n-1}}$ seen as an $\mathbb{F}_{q^2}$-linear transformation of $\mathbb{F}_{q^{2n}}$ induces a Singer cycle s for $\mathrm{GU}_n(q)$, with respect to a suitable choice of Hermitian form. Note that, by (2.3),

$$\det(s) = \left(\det{}_{\mathbb{F}_{q^2}}(\pi_a)\right)^{q^n-1} = \left(a^{\frac{q^{2n}-1}{q^2-1}}\right)^{q^n-1} = a^{\frac{(q^{2n}-1)(q^n-1)}{q^2-1}} \in \mathbb{F}_{q^2}^*$$

belongs to the subgroup of order $q+1$ of $\mathbb{F}_{q^2}^*$.

For the symplectic and orthogonal case, with n even, let $\langle a \rangle = \mathbb{F}_{q^n}^*$. Then, by [57, Theorem 5.6], the multiplication $\pi_{a^{q^{n/2}-1}}$ seen as an $\mathbb{F}_q$-linear transformation of V is a Singer cycle s both for $\mathrm{Sp}_n(q)$ and for $\mathrm{O}_n^-(q)$ with

$$\det(s) = \left(a^{\frac{q^n-1}{q-1}}\right)^{q^{n/2}-1} = 1.$$

In particular $s \in \mathrm{SO}_n^-(q)$.

In Table 2.1 we report the general classical groups $\hat{G}$ admitting Singer cycles, the order of the Singer cycles in $\hat{G}$ and the order of the Singer cycles in $\check{G}$ for every

2.6 Some Arithmetics

Table 2.1 Singer cycles

$\hat{G}$	Order of Singer cycle in $\hat{G}$	Order of Singer cycle in $\tilde{G}$	Comments
$GL_n(q)$	$q^n - 1$	$\frac{q^n-1}{q-1}$	
$GU_n(q)$	$q^n + 1$	$\frac{q^n+1}{q+1}$	n odd
			$\tilde{G} \neq SU_3(2)$
$Sp_n(q)$	$q^{\frac{n}{2}} + 1$	$q^{\frac{n}{2}} + 1$	
$O_n^-(q)$	$q^{\frac{n}{2}} + 1$	$\frac{q^{\frac{n}{2}}+1}{\gcd(2,q-1)}$	$\tilde{G} \neq \Omega_2^-(3)$

$n \geq 1$ with n odd in the unitary case and n even in the symplectic and orthogonal case. While the information in the second column comes directly from [60] and no exception for the existence arises, the third column needs a little explanation and two true exceptions arise as clarified in Lemma 2.4 and Proposition 2.5.

2.6 Some Arithmetics

The following lemma presents some arithmetical facts which will be very useful throughout the book. This is possibly part of folklore but we could not find a complete reference.

Lemma 2.2 *Let $q \geq 2$ be an integer. For every a and b positive integers we have*

(1) $\gcd(q^a - 1, q^b - 1) = q^{\gcd(a,b)} - 1$,

(2)
$$\gcd(q^a + 1, q^b + 1) = \begin{cases} q^{\gcd(a,b)} + 1 & \text{if } \frac{a}{\gcd(a,b)} \text{ and } \frac{b}{\gcd(a,b)} \text{ are odd,} \\ \gcd(2, q-1) & \text{otherwise,} \end{cases}$$

(3)
$$\gcd(q^a + 1, q^b - 1) = \begin{cases} q^{\gcd(a,b)} + 1 & \text{if } \frac{b}{\gcd(a,b)} \text{ is even,} \\ \gcd(2, q-1) & \text{otherwise.} \end{cases}$$

Proof The proof of all the formulas (1)–(3) is by induction on $a + b$. If $a + b = 2$, then we have $a = b = 1$ and all the formulas are clear. Now, let n be a positive integer with $n \geq 2$. Assume that (1)–(3) hold for every a and b with $a + b \leq n$ and we show that they hold also for every a and b with $a + b = n + 1$.

(1) The formula is symmetric in a and b and trivially holds for $a = b$. Thus we assume that $a > b$. Since

$$q^a - 1 = q^{a-b}(q^b - 1) + (q^{a-b} - 1),$$

by the inductive hypothesis for (1) we deduce that

$$\gcd(q^a - 1, q^b - 1) = \gcd(q^b - 1, q^{a-b} - 1) = q^{\gcd(a,b)} - 1,$$

because $a + (b - a) = b < a + b$.

(2) The formula is symmetric in a and b and trivially holds for $a = b$. Thus we assume that $a > b$. Since

$$q^a + 1 = q^{a-b}(q^b + 1) - (q^{a-b} - 1),$$

by the inductive hypothesis for (3) we deduce that

$$\gcd(q^a + 1, q^b + 1) = \gcd(q^b + 1, q^{a-b} - 1)$$

$$= \begin{cases} q^{\gcd(a,b-a)} + 1 & \text{when } \frac{a-b}{\gcd(a,a-b)} \text{ is even,} \\ \gcd(2, q - 1) & \text{otherwise.} \end{cases}$$

Now it is enough to observe that $\gcd(a, a - b) = \gcd(a, b)$ and $(a - b)/\gcd(a, b - a)$ is even if and only if both $a/\gcd(a, b)$ and $b/\gcd(a, b)$ are odd.

(3) In this case, the formula is not symmetric in a and b and hence we need to distinguish three cases: $a = b$, $a < b$ and $a > b$. When $a = b$, the formula can be easily checked. Assume next that $a < b$. Since

$$q^b - 1 = q^{b-a}(q^a + 1) - (q^{b-a} + 1),$$

by the inductive hypothesis for (2) we deduce that

$$\gcd(q^a + 1, q^b - 1) = \gcd(q^a + 1, q^{b-a} + 1)$$

$$= \begin{cases} q^{\gcd(a,b-a)} + 1 & \text{when } \frac{a}{\gcd(a,b-a)} \text{ and } \\ & \frac{b-a}{\gcd(a,b-a)} \text{ are odd,} \\ \gcd(2, q - 1) & \text{otherwise.} \end{cases}$$

Now it is enough to observe that $\gcd(a, a - b) = \gcd(a, b)$ and that $\frac{a}{\gcd(a,b)}$, $\frac{b-a}{\gcd(a,b)}$ are both odd if and only if $\frac{b}{\gcd(a,b)}$ is even.

Assume finally that $a > b$. Since

$$q^a + 1 = q^{a-b}(q^b - 1) + (q^{a-b} + 1),$$

by the inductive hypothesis for (3) we deduce that

$$\gcd(q^a + 1, q^b - 1) = \gcd(q^b - 1, q^{a-b} + 1)$$
$$= \begin{cases} q^{\gcd(a,b)} + 1 & \text{when } \frac{b}{\gcd(a,b)} \text{ is even,} \\ \gcd(2, q-1) & \text{otherwise.} \end{cases}$$

□

Given a positive integer n, we denote by $\pi(n)$ the set of prime divisors of n.

2.7 Primitive Prime Divisors

Given a prime power q and an integer $t \geq 2$, a prime r is called a ***primitive prime divisor*** of $q^t - 1$ if r divides $q^t - 1$ and r does not divide $q^i - 1$, for each $i \in \{1, \ldots, t-1\}$. We let $P_t(q)$ denote *the set of primitive prime divisors* of $q^t - 1$. From a celebrated theorem of Zsigmondy [94], the following hold

- for $t \geq 3$, we have $P_t(q) \neq \emptyset$ with the only exception of $(t, q) = (6, 2)$,
- $P_2(q) \neq \emptyset$ with the only exception of q being a Mersenne prime.

Note that if r is a primitive prime divisor of $q^t - 1$, then q has order t modulo r and thus t divides $r - 1$. Hence there exists a positive integer k such that $r = kt + 1$. In particular,

$$r \geq t + 1, \tag{2.4}$$

and when t is odd

$$r \geq 2t + 1. \tag{2.5}$$

We collect in the following lemma some elementary facts about $P_t(q)$.

Lemma 2.3 *Let s, t, k, b be positive integers with $s, t \geq 2$ and let q be a prime power. Then the following hold*

(1) $P_{kt}(q) \subseteq P_t(q^k)$,
(2) *if $t \neq s$, then $P_t(q) \cap P_s(q) = \emptyset$,*
(3) *given $r \in P_t(q)$, r divides $q^b - 1$ if and only if t divides b. In particular, if r divides $q^b + 1$, then t divides $2b$.*

2.8 Further Facts on Singer Cycles

The concept of primitive prime divisor allows one, among other things, to obtain an easy test to recognize the Singer cycles by checking only the element orders. We first treat the action of $\Omega_2^-(q)$ on its natural module V.

Lemma 2.4 *The cyclic group $\Omega_2^-(q) \cong C_{\frac{q+1}{(2,q-1)}}$ acts irreducibly on V if $q \neq 3$ and scalarly if $q = 3$. In particular, $\Omega_2^-(3)$ admits no Singer cycle.*

Proof This follows from [66, Section 2.10], but here we give a direct proof. If V is not an irreducible $\mathbb{F}_q \Omega_2^-(q)$-module, then V decomposes as the direct sum of two 1-dimensional $\mathbb{F}_q \Omega_2^-(q)$-submodules. Hence $\Omega_2^-(q)$ is conjugate to a cyclic subgroup of the group of 2×2 diagonal matrices. Therefore $(q+1)/\gcd(2, q-1) = |\Omega_2^-(q)|$ divides $q - 1$. By Lemma 2.2, this yields $q = 3$. On the other hand, $\Omega_2^-(3) \cong C_2$ is the group of scalar matrices of $\mathrm{GL}_2(3)$ so that obviously $\Omega_2^-(3)$ cannot admit cyclic subgroups acting irreducibly on V. □

In particular, Lemma 2.4 explains the exception of $\Omega_2^-(3)$ in Table 2.1.

Proposition 2.5 *Let $\hat{G}$ be a general classical group and $\tilde{G}$ be the corresponding classical group. Then the following hold:*

(1) *Assume that $\hat{G}$ or $\tilde{G}$ are as in Table 2.1. If x belongs to $\hat{G}$ or to $\tilde{G}$ and has the order of a Singer cycle in $\hat{G}$ or $\tilde{G}$ respectively, then x is a Singer cycle.*
(2) *$\tilde{G}$ admits a Singer cycle if and only if it appears in Table 2.1.*
(3) *Let $\hat{G} = \mathrm{GU}_n(q)$, with n odd and $\lambda \in \mathbb{F}_{q^2}^*$ be an element of order $q + 1$. Then there exists a Singer cycle s of $\hat{G}$ such that $\det(s) = \lambda$.*
(4) *Let $\hat{G} = \mathrm{GL}_n(q)$, and $\lambda \in \mathbb{F}_q^*$ be an element of order $q - 1$. Then there exists a Singer cycle s of $\hat{G}$ such that $\det(s) = \lambda$.*

Proof Let V be the natural n-dimensional $\mathbb{F}_{q^\delta} \hat{G}$-module. If $n = 1$ all the statements are trivial. So assume that $n \geq 2$.

(1) Regard V as an $\mathbb{F}_{q^\delta} \langle x \rangle$-module and assume, by contradiction, that V is not irreducible. Then there exists an irreducible $\mathbb{F}_{q^\delta} \langle x \rangle$-submodule W of V and $m = \dim_{\mathbb{F}_{q^\delta}} W$ is a positive integer less than n. By Schur's Lemma, we have that $\langle x \rangle \leq C_{q^{\delta m}-1}$ and hence

$$\mathrm{o}(x) \mid q^{\delta m} - 1. \tag{2.6}$$

If $\hat{G} = \mathrm{GL}_n(q)$, then (2.6) becomes $q^n - 1 \mid q^m - 1$, against $q^m - 1 < q^n - 1$.
If $\tilde{G} = \mathrm{SL}_n(q)$, then (2.6) means $\frac{q^n-1}{q-1} \mid q^m - 1$. Assume first that $(n, q) \neq (6, 2)$ and $(n, q) \neq (2, p)$ for all the Mersenne primes p. Then there exists $r \in P_n(q)$. Since $r \nmid q - 1$, we have that $r \mid \frac{q^n-1}{q-1}$ and $\frac{q^n-1}{q-1} \mid q^m - 1$; however, putting together these divisibility conditions we obtain a contradiction. If $(n, q) = (6, 2)$, then $\mathrm{SL}_n(q) = \mathrm{SL}_6(2) = \mathrm{GL}_6(2)$ and

2.8 Further Facts on Singer Cycles

we have seen before that the result holds. Let next $(n, q) = (2, p)$ for some Mersenne prime p. Then $m = 1$ and (2.6) becomes $p + 1 \mid p - 1$, which is obviously impossible.

If $\hat{G} = \mathrm{GU}_n(q)$, for some $n \geq 3$ odd, then (2.6) becomes $q^n + 1 \mid q^{2m} - 1$. Assume first that $(n, q) \neq (3, 2)$. Then $(2n, q) \neq (6, 2)$ and $2n \neq 2$ so that there exists $r \in P_{2n}(q)$. Since $r \mid q^n + 1$ and $q^n + 1 \mid q^{2m} - 1$ with $2m < 2n$, we get a contradiction. When $(n, q) = (3, 2)$ we have $\hat{G} = \mathrm{GU}_3(2) \leq \mathrm{GL}_3(4)$, $\mathbf{o}(x) = 9$ and (2.6) becomes $9 \mid 4^m - 1$, which is impossible for all $1 \leq m \leq 2$.

If $\tilde{G} = \mathrm{SU}_n(q)$, for some $n \geq 3$ odd, then (2.6) becomes $\frac{q^n+1}{q+1} \mid q^{2m} - 1$. Assume first $(n, q) \neq (3, 2)$. Pick $r \in P_{2n}(q) \neq \emptyset$ and note that $r \nmid q^2 - 1$ so that $r \nmid q + 1$. Since $r \mid q^n + 1$, we also have that $r \mid \frac{q^n+1}{q+1} \mid q^{2m} - 1$, a contradiction. When $(n, q) = (3, 2)$ we have $\tilde{G} = \mathrm{SU}_3(2)$ and $\mathbf{o}(s) = 3$. This is a true exception because $\mathrm{SU}_3(2)$ does not admit Singer cycles. Assume, on the contrary, that y is a Singer cycle of $\mathrm{SU}_3(2)$ and let v be an arbitrary non-zero vector in $V = \mathbb{F}_4^3$. Set $w := v + vy + vy^2$. Observe that $wy = vy + vy^2 + vy^3 = w$. If $w \neq 0$, this shows that y stabilizes the 1-dimensional subspace of V spanned by w, contradicting the fact that y acts irreducibly. If $w = 0$, then $vy^2 = v + vy \in \langle v, vy \rangle$ and hence y stabilizes the non-zero proper subspace $\langle v, vy \rangle$ of V, contradicting the fact that y acts irreducibly. Therefore, $\mathrm{SU}_3(2)$ has no Singer cycles and hence it is one of the exceptions in Table 2.1.[1]

Let next $\hat{G} = \tilde{G} = \mathrm{Sp}_n(q)$. If $n = 2$, then (2.6) becomes $q + 1 \mid q - 1$, which is absurd. Let then $n \geq 4$ be even. Then (2.6) becomes $q^{n/2} + 1 \mid q^m - 1$. Assume first that $(n, q) \neq (6, 2)$. Then, since $n \neq 2$, there exists $r \in P_n(q)$ and $r \mid q^{n/2} + 1 \mid q^m - 1$, a contradiction. Let now $(n, q) = (6, 2)$, so that $\tilde{G} = \mathrm{Sp}_6(2)$. Then (2.6) becomes $9 \mid 2^m - 1$, which is impossible for all $1 \leq m \leq 5$.

Let now $\hat{G} = \mathrm{O}_n^-(q)$ with n even. Then (2.6) becomes $q^{n/2} + 1 \mid q^m - 1$ and the same arithmetic arguments used for the symplectic case apply.

Let us finally consider $\tilde{G} = \Omega_n^-(q)$ with n even and $(n, q) \neq (2, 3)$ as required in Table 2.1. The possibility $n = 2$ cannot arise because when $n = 2$ we have $\mathbf{o}(x) = \frac{q+1}{\gcd(2, q-1)}$ so that $\langle x \rangle = \Omega_2^-(q)$ and, by Lemma 2.4, V is irreducible. Assume then $n \geq 4$. Since we have observed that $9 \mid 2^m - 1$ is impossible for all $1 \leq m \leq 5$, we are left with $(n, q) \neq (6, 2)$. Thus $P_n(q) \neq \emptyset$ and we argue as in the symplectic case because, by (2.4), any $r \in P_n(q)$ is odd and hence divides $\frac{q^{n/2}+1}{\gcd(2, q-1)}$.

(2) Assume that $\tilde{G}$ appears in Table 2.1. By (1) it is enough to exhibit an element of $\tilde{G}$ having the order of a Singer cycle of $\tilde{G}$. This is immediately done by suitable multiplications. Indeed, let $\langle a \rangle = \mathbb{F}_{q^{\delta n}}^*$. By what is explained in Sect. 2.5 we

---

[1] Singer elements in $\mathrm{SU}_3(q)$ are discussed in detail in [5, Section 2]. However, there is a slight ambiguity in the discussion on $\mathrm{SU}_3(2)$ there and hence, rather than referencing directly to [5], we have included our own argument.

have that $\pi_{a^{q-1}} \in \mathrm{SL}_n(q)$ and has order $\frac{q^n-1}{q-1}$; for n odd, $\pi_{a^{(q^n-1)(q+1)}} \in \mathrm{SU}_n(q)$ and has order $\frac{q^n+1}{q+1}$; $\pi_{a^{q^{n/2}-1}} \in \mathrm{Sp}_n(q)$ and has order $q^{n/2}+1$; $\pi_{a^{2(q^{n/2}-1)}} \in \Omega_n^-(q)$ and has order $\frac{q^{\frac{n}{2}}+1}{\gcd(2,q-1)}$.

Conversely, assume that $\tilde{G}$ admits a Singer cycle s. Then $\tilde{G} \neq \Omega_2^-(3)$ because $\Omega_2^-(3)$ does not admit Singer cycles, by Lemma 2.4. Moreover, $\tilde{G} \neq \mathrm{SU}_3(2)$, because above we have proved that $\mathrm{SU}_3(2)$ does not admit Singer cycles. Observe that $\langle s \rangle$ is a cyclic subgroup of $\hat{G}$ acting irreducibly on V and thus, by the results in [60], the possible groups $\hat{G}$ are those appearing in Table 2.1. But then also $\tilde{G}$ appear in Table 2.1.

(3) Let $a \in \mathbb{F}_{q^{2n}}^*$ be such that $\langle a \rangle = \mathbb{F}_{q^{2n}}^*$ and look at π_a as an $\mathbb{F}_{q^2}$-linear transformation of $\mathbb{F}_{q^{2n}}$. By (2.3), we have $\det_{\mathbb{F}_{q^2}}(\pi_a) = a^{\frac{q^{2n}-1}{q^2-1}} \in \mathbb{F}_{q^2}^*$ so that $\mathrm{o}(\det_{\mathbb{F}_{q^2}}(\pi_a)) = q^2 - 1$. By Sect. 2.5, we know that $s := \pi_{a^{q^n-1}}$ is a Singer cycle for $\mathrm{GU}_n(q)$. Let $\mu := \det(s) = (\det_{\mathbb{F}_{q^2}}(\pi_a))^{q^n-1} \in \mathbb{F}_{q^2}^*$ and observe that, by Lemma 2.2, we have

$$\mathrm{o}(\mu) = \frac{q^2-1}{\gcd(q^2-1, q^n-1)} = \frac{q^2-1}{q-1} = q+1,$$

so that μ generates the subgroup U of $\mathbb{F}_{q^2}^*$ of size $q+1$. Since λ also generates U, there exists an integer k with $1 \leq k \leq q+1$ and $\gcd(k, q+1) = 1$ such that $\lambda = \mu^k$. Consider now the arithmetic progression $k_\ell := k + \ell(q+1)$. By Dirichlet's Theorem there exists ℓ' such that $k' := k + \ell'(q+1)$ is a prime number and $k' > q^n + 1$. Define then $s' := s^{k'}$. Since k' is coprime with $\mathrm{o}(s) = q^n + 1$, we have $\langle s' \rangle = \langle s \rangle$ and hence s' is a Singer cycle for $\mathrm{GU}_n(q)$. Moreover $\det(s') = \mu^{k'} = \mu^k = \lambda$.

(4) The proof is similar to that of (3) and thus omitted.

$\square$

2.9 The *ppd*-Elements

We introduce now the fundamental facts about ***primitive prime divisor elements*** (a.k.a. *ppd*-elements) developed by Guralnick et al. in [53]. Throughout this book, that theory is the main tool in finding the maximal subgroups containing elements with order divisible by certain "large" primes in classical groups.

An element $x \in \mathrm{GL}_n(q)$ is said to be a *ppd*$(n, q; e)$-**element**, for e an integer such that $n/2 < e \leq n$, if $\mathrm{o}(x)$ is divisible by some prime $r \in P_e(q)$. Note that this notion implicitly requires that $P_e(q) \neq \emptyset$.

A subgroup M of $\mathrm{GL}_n(q)$ containing a *ppd*$(n, q; e)$-element is said a *ppd*$(n, q; e)$-**group**. The main result in [53] gives a satisfactory description of the

2.9 The *ppd*-Elements

$ppd(n,q;e)$-groups. There the $ppd(n,q;e)$-groups are divided into nine classes, described through the Examples 2.1–2.9, and it is shown that every $ppd(n,q;e)$-group belongs to one of these classes. We are particularly interested in maximal $ppd(n,q;e)$-groups. In particular, with a careful analysis of the Examples 2.1–2.9 from [53], we may deduce the following result.

Theorem 2.6 ([53, Main Theorem]) *Let $q = p^f$ be a prime power and let n be an integer with $n \geq 2$. Then a subgroup M of $\mathrm{GL}_n(q)$ is a $ppd(n,q;e)$-group if and only if M is one of the groups in Examples 2.1–2.9 in [53]. Moreover,*

(1) $M \notin \mathcal{C}_4 \cup \mathcal{C}_7$;
(2) *if $M \in \mathcal{S}$, then M is one of the groups described in Examples 2.6–2.9;*
(3) *if $e \leq n - 3$ and M is one of the groups described in Examples 2.5, 2.6 b), 2.6 c), 2.7, 2.8, 2.9, then using the notation in [53] one of the followings holds:*

- $n = 7$, $e = 4$, $M' \cong \mathrm{Sp}_6(2)$ and $p > 2$,
- $n = 9$, $e = 6$, $M' \cong \mathrm{SL}_3(q)$, f is even and $\sqrt{q} \equiv 1 \pmod{3}$,
- $n = 9$, $e = 6$, $M' \cong \mathrm{PSL}_3(q)$, f is even and $\sqrt{q} \not\equiv 1 \pmod{3}$,
- $n = s + 1$, $e = s - 2$, $M' \cong \mathrm{PSL}_2(s)$, $s \geq 7$, $s = 2^c$ with c a prime.

Moreover, when $e \leq n - 4$, M does not lie in $\mathcal{C}_6$ and if M does lie in $\mathcal{S}$, then M appears in Example 2.6 a) of [53].

Observe that, when $e \leq n - 3$ and M is one of the groups described in Examples 2.6a), Theorem 2.6 does not give any additional information. This lack of additional information will play little role in our proofs later.

In order to exclude the Aschbacher class $\mathcal{C}_5$, it is useful to introduce a further notion. A $ppd(n,q;e)$-element is called a **strong $ppd(n,q;e)$-element** if its order is divisible by every $r \in P_e(q)$. Of course also this notion implicitly requires that $P_e(q) \neq \emptyset$.

Lemma 2.7 *Let $\tilde{G} \leq \mathrm{GL}_n(q^\delta)$ be a classical group and let $n/2 < e \leq n$. Then the following hold*

(1) *If $M \in \mathcal{C}_5$ is a maximal subgroup of $\tilde{G}$, then there exists a prime $k \mid \delta f$ such that $M \leq \mathrm{GL}_n(q^{\delta/k}) = \mathrm{GL}_n(p^{(\delta f)/k})$.*
(2) *Let $M \in \mathcal{C}_5$ be a maximal subgroup of $\tilde{G}$ with $M \leq \mathrm{GL}_n(q^{\delta/k})$, where $k \mid \delta f$ is a prime. Then M contains no element of $\tilde{G}$ having order divisible by some prime in $P_{ke}(q^{\delta/k})$. Moreover, if $P_{ke}(q^{\delta/k}) \neq \emptyset$, then M contains no strong $ppd(n,q;e)$-element.*
(3) *If $n \geq 6$, then a maximal subgroup of $\tilde{G}$ in the class $\mathcal{C}_5$ never contains a strong $ppd(n,q;e)$-element.*

Proof

(1) This follows immediately by the definition of the maximal subgroups M of $\tilde{G}$ belonging to class $\mathcal{C}_5$ as given in [66, Chapters 3 and 4]. Note that in the unitary case $\delta = 2$ and there are three cases to consider: one with k odd and two with $k = 2$.

(2) Assume, by contradiction, that there exist $y \in \tilde{G}$ and $r \in P_{ke}(q^{\delta/k})$ such that $r \mid \mathbf{o}(y)$ and $y \in M$. Then we have

$$p \neq r \mid |M| \mid |\mathrm{GL}_n(q^{\delta/k})|.$$

Hence $r \mid (q^{\delta/k})^i - 1$ for some $i \in \{1, \ldots, n\}$. By the definition of primitive prime divisor, this implies $ke \leq n$. Then we get $n \geq ke \geq 2e > n$, a contradiction.

Assume next that M contains a strong $ppd(n, q; e)$-element $y \in \tilde{G}$. Using Lemma 2.3 (1), we have $P_{ke}(q^{\delta/k}) \subseteq P_e(q^\delta)$. Since every prime in $P_e(q^\delta)$ divides $\mathbf{o}(y)$ and since $P_{ke}(q^{\delta/k}) \neq \emptyset$, there exists $r \in P_{ke}(q^{\delta/k})$ such that $r \mid \mathbf{o}(y)$, in contrast to what was previously shown.

(3) When $n \geq 6$, we have $ke \geq 2e \geq n + 1 \geq 7 > 6$ and thus, by Zsigmondy's theorem, $P_{ke}(q^{\delta/k}) \neq \emptyset$. Now apply (2). □

2.10 Bertrand Elements

We use $ppd(d, q; e)$-elements only for some special values t of e, which we now describe. Recall that, for every $n \geq 7$, Bertrand's postulate guarantees the existence of a prime number t such that $n/2 < t \leq n - 2$. When $n \geq 8$, it follows from Bertrand's postulate that there exists a prime t such that $n/2 < t \leq n - 3$, see [55, p. 273]. The relevance of $n - 3$ here is related to $n - 3$ appearing in Theorem 2.6 part (3), but this will be more clear when we will embark in the proof of Theorems 1.3 and 1.5. We call such a prime t a **Bertrand number** for n. We extend the notion of Bertrand numbers also when $n = 5, 6, 7$ by setting $t := 3, 4, 5$, respectively. We do not define Bertrand numbers when $n < 5$. Note that if t is a Bertrand number for $n \geq 5$, then $t \nmid n$ and $n/2 < t \leq n - 2$.

In Table 2.2, we define an element z in some classical groups depending on a certain Bertrand number t. We refer to such a z as a **Bertrand element**. In this preliminary section, we only give a very rough description of z. More details are given inside each section dealing with a certain family of classical groups. In particular, the existence of the Bertrand elements and their action on V will be specified in due course. Strictly speaking, in the second column of Table 2.2 we only

Table 2.2 Bertrand elements z

G	Order of z	t	e
$\mathrm{SU}_n(q)$, $n \neq 6$	$\dfrac{(q^t+1)(q^{n-t}+(-1)^n)}{\gcd(q^t+1, q^{n-t}+(-1)^n)}$	t Bertrand number for n	t
$\mathrm{Sp}_n(q)$	$\dfrac{(q^t+1)(q^{n/2-t}+1)}{\gcd(q^t+1, q^{n/2-t}+1)}$	t Bertrand number for $\frac{n}{2}$	$2t$
$\Omega_n(q)$	$\dfrac{(q^t+1)(q^{(n-1)/2-t}+1)}{\gcd(q^t+1, q^{(n-1)/2-t}+1)}$	t Bertrand number for $\frac{n-1}{2}$	$2t$

define the order of z. In the third column of Table 2.2 we define the natural number with respect to which t is a Bertrand number. Apart from a few degenerate cases, Bertrand elements are $ppd(n, q; e)$-elements and we define in the fourth column of Table 2.2 the value of e. The degenerate cases will be discussed later.

2.11 The Spinor Norm and the Bertrand Elements

We recall the definition and properties of the spinor norm following the description in [66, pp. 29, 30]. Let $\varepsilon \in \{+, -\}$, let ℓ be an even positive integer and q be odd. Consider the special orthogonal group $\mathrm{SO}_\ell^\varepsilon(q)$ with respect to a non-degenerate quadratic form $Q : \mathbb{F}_q^\ell \to \mathbb{F}_q$ on $\mathbb{F}_q^\ell$. We let

$$\langle \cdot, \cdot \rangle : \mathbb{F}_q^\ell \times \mathbb{F}_q^\ell \to \mathbb{F}_q$$

be the non-degenerate symmetric form polarizing to Q. This is clearly well-defined and unique because q is odd. Since q is odd, we also infer that the multiplicative group $\mathbb{F}_q^*$ of $\mathbb{F}_q$ has even order $q - 1$ and hence the subgroup

$$(\mathbb{F}_q^*)^2 := \{x^2 \mid x \in \mathbb{F}_q^*\}$$

of $\mathbb{F}_q^*$ consisting of the square elements of the field $\mathbb{F}_q$ different from 0 has index 2 in $\mathbb{F}_q^*$. In other words, $\mathbb{F}_q^*/(\mathbb{F}_q^*)^2$ is a cyclic group of order 2. Let $\lambda \in \mathbb{F}_q^* \setminus (\mathbb{F}_q^*)^2$. Then $\langle \lambda (\mathbb{F}_q^*)^2 \rangle = \mathbb{F}_q^*/(\mathbb{F}_q^*)^2$ and $\lambda^2 \in (\mathbb{F}_q^*)^2$.

From [66, Proposition 2.5.6], every element $g \in \mathrm{SO}_\ell^\varepsilon(q)$ can be written as the product of an even number of reflections, that is, $g = r_{v_1} \cdots r_{v_{2\kappa}}$ for some reflections $r_{v_1}, \ldots, r_{v_{2\kappa}}$ and for some non-negative integer κ. Here, following the notation in [66], we are denoting with r_v the reflection with respect to the axis $v \in \mathbb{F}_q^\ell$. Observe that the reflection

$$x \mapsto xr_v = x - \frac{\langle x, v \rangle}{Q(v)} v$$

is well-defined only when v is a non-isotropic vector with respect to Q. The spinor norm is the mapping

$$\theta : \mathrm{SO}_\ell^\varepsilon(q) \to \mathbb{F}_q^*/(\mathbb{F}_q^*)^2$$

defined by

$$g = r_{v_1} \cdots r_{v_{2\kappa}} \mapsto \prod_{i=1}^{2\kappa} \langle v_i, v_i \rangle \pmod{(\mathbb{F}_q^*)^2}.$$

Observe that since v_i is non-isotropic for each i, we have $\prod_{i=1}^{2\kappa}\langle v_i, v_i\rangle \neq 0$ and hence $\prod_{i=1}^{2\kappa}\langle v_i, v_i\rangle \in \mathbb{F}_q^*$. There are various things remarkable about this mapping. First, θ is well defined, that is, the value of $\theta(g)$ does not depend on the way we express g as a product of reflections. Second, θ is a group homomorphism from the special orthogonal group $\mathrm{SO}_\ell^\varepsilon(q)$ to the cyclic group $\mathbb{F}_q^*/(\mathbb{F}_q^*)^2$ of order 2. Third, θ is surjective. Fourth, the kernel of θ is a subgroup of $\mathrm{SO}_\ell^\varepsilon(q)$ having index 2 and it is indeed our player $\Omega_\ell^\varepsilon(q)$. All of these facts can be found in [66, pp. 29, 30] and in the bibliography therein.

For every q, since $|\mathrm{SO}_\ell^\varepsilon(q) : \Omega_\ell^\varepsilon(q)|$ is a power of 2, we deduce that $\Omega_\ell^\varepsilon(q)$ contains all the odd order elements of $\mathrm{SO}_\ell^\varepsilon(q)$.

The situation when q is even is much simpler.[2] Indeed, except for $\mathrm{O}_4^+(2)$, when q is even, $\Omega_\ell^\varepsilon(q)$ is the subgroup of $\mathrm{SO}_\ell^\varepsilon(q)$ consisting of products of an even number of reflections. For more details see [66, p. 30] and [66, Proposition 2.5.9].

We now present a result which, among other things, will help us in proving the existence of Bertrand elements for the orthogonal groups throughout the book.

Proposition 2.8 *Let n, m be even positive integers with $2 \leq m \leq n/2$ and consider the embedding of $\mathrm{SO}_m^-(q) \perp \mathrm{SO}_{n-m}^-(q)$ in $\mathrm{SO}_n^+(q)$. Let $s_m \in \mathrm{SO}_m^-(q)$ and $s_{n-m} \in \mathrm{SO}_{n-m}^-(q)$ be Singer cycles and define $x := s_m \oplus s_{n-m} \in \mathrm{SO}_n^+(q)$. Then x has action type $m \oplus (n-m)$,*

$$\mathbf{o}(x) = \frac{(q^{\frac{m}{2}}+1)(q^{\frac{n}{2}-\frac{m}{2}}+1)}{\gcd(q^{\frac{m}{2}}+1, q^{\frac{n}{2}-\frac{m}{2}}+1)}$$

and $x \in \Omega_n^+(q)$.

Proof We have observed in Sect. 2.5 that $\mathrm{SO}_m^-(q)$ and $\mathrm{SO}_{n-m}^-(q)$ contain Singer cycles s_m and s_{n-m} having order $q^{m/2}+1$ and $q^{n/2-m/2}+1$, respectively. Thus $\mathbf{o}(x)$ and its action immediately follows and we only need to show that $x \in \Omega_n^+(q)$. From [66, Table 2.1C], we have $|\mathrm{SO}_n^+(q) : \Omega_n^+(q)| = 2$. If q is even, we immediately get $x \in \Omega_n^+(q)$, because $\mathbf{o}(x)$ is odd. Suppose next that q is odd. Let $\theta : \mathrm{SO}_n^+(q) \to \mathbb{F}_q^*/(\mathbb{F}_q^*)^2$ be the spinor norm of $\mathrm{SO}_n^+(q)$. Observe that the spinor norms of $\mathrm{SO}_m^-(q)$ and $\mathrm{SO}_{n-m}^-(q)$ are simply the restrictions of the spinor norm θ to $\mathrm{SO}_m^-(q)$ and $\mathrm{SO}_{n-m}^-(q)$, respectively, because the quadratic forms defining $\mathrm{SO}_m^-(q)$ and $\mathrm{SO}_{n-m}^-(q)$ are simply the restrictions of the quadratic form defining $\mathrm{SO}_n^+(q)$.

From Table 2.1, we see that, for every even positive integer ℓ, when q is odd, a Singer cycle in $\mathrm{O}_\ell^-(q)$ is never contained in $\Omega_\ell^-(q)$. Hence $\Omega_m^-(q)$ and $\Omega_{n-m}^-(q)$ do not contain s_m and s_{n-m} and hence the spinor norms of s_m and s_{n-m} are not the identity, that is,

$$\theta(s_m) = \lambda(\mathbb{F}_q^*)^2 \quad \text{and} \quad \theta(s_{n-m}) = \lambda(\mathbb{F}_q^*)^2,$$

where $\lambda \in \mathbb{F}_q^*$ and $\langle\lambda(\mathbb{F}_q^*)^2\rangle = \mathbb{F}_q^*/(\mathbb{F}_q^*)^2$.

---

[2] Reflections can be defined in an entirely similar way as the case q odd.

2.11 The Spinor Norm and the Bertrand Elements

It follows that

$$\theta(x) = \theta(s_m \oplus s_{n-m}) = \theta(s_m)\theta(s_{n-m}) = \lambda^2(\mathbb{F}_q^*)^2 = (\mathbb{F}_q^*)^2.$$

This shows that $\theta(x)$ is the identity element of $\mathbb{F}_q^*/(\mathbb{F}_q^*)^2$ and hence $x \in \Omega_n^+(q)$. □

Sometimes, throughout the book, we will freely use some of the facts in this chapter with no further reference.

Chapter 3
Linear Groups

In Sect. 1.8, we have explained that, by [14], we know that $\gamma_w(\mathrm{PSL}_n(q)) \geq 3$ for $n \geq 5$ and $\gamma_w(\mathrm{PSL}_n(q)) \leq 2$ for $2 \leq n \leq 4$. Thus we only need to deal with projective special linear groups $\mathrm{PSL}_n(q)$ with $2 \leq n \leq 4$, where $q \notin \{2, 3\}$ when $n = 2$, with the purpose to exclude $\gamma_w(\mathrm{PSL}_n(q)) = 1$ and to describe the maximal components of a weak normal 2-covering.

Since $\mathrm{PSL}_2(4) \cong \mathrm{PSL}_2(5) \cong A_5$ and $\mathrm{PSL}_2(9) \cong A_6$ and since we have already discussed alternating groups in Sect. 1.8, we may suppose that, when $n = 2$, we have $q \notin \{2, 3, 4, 5, 9\}$.

Recall that, given a subgroup X of $\mathrm{PSL}_n(q)$, we denote by $\tilde{X}$ its preimage under the natural projection $\mathrm{SL}_n(q) \to \mathrm{PSL}_n(q)$. Moreover, since the automorphism group $\mathrm{Aut}(\mathrm{SL}_n(q))$ projects onto the automorphism group of $\mathrm{PSL}_n(q)$, in the proofs of the lemmas in this section we may work with the linear group $\mathrm{SL}_n(q)$ (see Sect. 2.2).

For some of the facts that we use in the proof of Lemmas 3.1, 3.2 and 3.3, we could simply refer to some results already known in the literature. However, we have decided to include full arguments of these lemmas because this will serve as a warm up for later.

In the proof of Lemma 3.1, we use [9, Tables 8.1 and 8.2] and hence we adopt the notation from [9]. In particular, given a positive even integer n, the group Q_{2n} has order $2n$ and defined by the presentation $\langle a, b \mid a^4 = 1, a^2 = b^{n/2}, a^{-1}ba = b^{-1}\rangle$. When $n = 4$, Q_8 is the usual quaternion group of order 8.

Lemma 3.1 *Let q be a prime power with $q \notin \{2, 3, 4, 5, 9\}$. Then $\gamma_w(\mathrm{PSL}_2(q)) = 2$. Moreover, if H and K are the two maximal components of a weak normal 2-covering of $\mathrm{PSL}_2(q)$, then up to $\mathrm{Aut}(\mathrm{PSL}_2(q))$-conjugacy one of the following holds*

(1) $q \neq 7$, $H \cong D_{2(q+1)/\gcd(2,q-1)}$ *is in class* $\mathcal{C}_3$ *and K is a parabolic subgroup in class* $\mathcal{C}_1$,

(2) q is even, $H \cong D_{2(q+1)}$ is in class $\mathcal{C}_3$ and $K \cong D_{2(q-1)}$ is in class $\mathcal{C}_2$,
(3) $q = 7$, $H \cong S_4$ is in class $\mathcal{C}_6$ and K is a parabolic subgroup in class $\mathcal{C}_1$.

Moreover, each of these weak normal 2-coverings gives rise to a single normal covering.

Proof We use the list of the maximal subgroups of $\mathrm{PSL}_2(q)$ in [9, Tables 8.1, 8.2] and the notation therein.

When $q \in \{7, 8, 11\}$, the proof follows with a computer computation. For the rest of the proof, we suppose that $q \geq 13$. Let μ be a weak normal covering of $\mathrm{PSL}_2(q)$ of minimum size and maximal components and $\tilde{\mu}$ be the corresponding weak normal covering of $\mathrm{SL}_2(q)$. Let x be an element of $\mathrm{SL}_2(q)$ having order $q+1$, that is, a Singer cycle for $\mathrm{SL}_2(q)$. From [9, Tables 8.1, 8.2] and from the fact that $q \notin \{5, 7, 9\}$, we see that the only maximal subgroups of $\mathrm{SL}_2(q)$ containing a Singer cycle form a unique $\mathrm{SL}_2(q)$-conjugacy class of subgroups given by the maximal subgroups in class $\mathcal{C}_3$. Therefore, without loss of generality, we may suppose that $x \in \tilde{H}$ for some $\tilde{H} \in \tilde{\mu}$ belonging to class $\mathcal{C}_3$.

From [9, Table 8.1, 8.2], we deduce $\tilde{H} \cong Q_{2(q+1)}$ when q is odd and $\tilde{H} \cong D_{2(q+1)}$ when q is even. In both cases, $\tilde{H} \cong (q+1).2$. Let y be an element of $\mathrm{SL}_2(q)$ having order $q-1$. Using the fact that $q \notin \{5, 7, 9, 11\}$, another case-by-case analysis on the maximal subgroups of $\mathrm{SL}_2(q)$ in [9, Tables 8.1, 8.2] reveals that there are two $\mathrm{Aut}(\mathrm{SL}_2(q))$-conjugacy classes of maximal subgroups containing an $\mathrm{Aut}(\mathrm{SL}_2(q))$-conjugate of y. Namely,

  (i) $E_q : (q-1) \in \mathcal{C}_1$,
  (ii) $Q_{2(q-1)} \in \mathcal{C}_2$, when q is odd,
  (iii) $D_{2(q-1)} \in \mathcal{C}_2$, when q is even.

As $\tilde{H}$ is not in this list, we deduce that there exists a further component $\tilde{K} \in \tilde{\mu}$. Thus $\gamma_w(\mathrm{PSL}_2(q)) = 2$ and $\tilde{K}$ must be in one of the above two possibilities.[1] Suppose first that $\tilde{K} = Q_{2(q-1)}$ when q is odd. Now, $\tilde{H}$ and $\tilde{K}$ do not contain unipotent elements and hence this case does not give rise to a weak normal 2-covering of $\mathrm{SL}_2(q)$. Therefore, when q is odd, the only possibility is $\tilde{K} = E_q : (q-1)$. Hence we obtain as candidates for the weak normal 2-coverings of $\mathrm{PSL}_2(q)$ those in (1) and (2).

We show that actually, in both cases, $\tilde{\mu} = \{\tilde{H}, \tilde{K}\}$ is a normal 2-covering of $\mathrm{SL}_2(q)$. Assume first that case (1) holds. Then $\tilde{H}$ contains a Singer cycle and $\tilde{K}$ is parabolic. Let $z \in \mathrm{SL}_2(q)$. If z has no eigenvalues over $\mathbb{F}_q$, then $\langle z \rangle$ acts irreducibly on the natural module $V = \mathbb{F}_q^2$ so that z is the power of a Singer cycle ([14, Lemma 2.4]). Since Singer subgroups of $\mathrm{SL}_2(q)$ are $\mathrm{SL}_2(q)$-conjugate, we deduce that z belongs to an $\mathrm{SL}_2(q)$-conjugate of $\tilde{H}$. If z admits an eigenvalue $\lambda \in \mathbb{F}_q^*$, then choosing a suitable eigenvector for λ and completing to a basis for V we construct

---

[1] When q is odd, $\tilde{K}$ is either $E_q : (q-1)$ or $Q_{2(q-1)}$ and, when q is even, $\tilde{K}$ is either $E_q : (q-1)$ or $Q_{2(q-1)}$.

3 Linear Groups

a matrix $c \in \mathrm{SL}_2(q)$ such that $c^{-1}zc$ is an upper triangular matrix with diagonal entries λ, λ^{-1} which belongs to $\tilde{K}$.

Assume next that case (2) holds. Thus q is even. Let $z \in \mathrm{SL}_2(q)$. If z has no eigenvalues, arguing as in case (1), we deduce that z belongs to an $\mathrm{SL}_2(q)$-conjugate of $\tilde{H}$. By [9, Lemma 3.1.3], we know the standard copy of $D_{2(q-1)}$ inside $\mathrm{SL}_2(q)$ and thus we can assume that $\tilde{K}$ is this standard copy. In other words, we can assume $\tilde{K} = \langle \mathrm{diag}(\omega, \omega^{-1}), \mathrm{antidiag}(1,1) \rangle$, where ω generates the multiplicative group $\mathbb{F}_q^*$. Suppose now that z has an eigenvalue $\lambda \in \mathbb{F}_q$. If $\lambda = 1$, then z is unipotent and, since involutions in $\mathrm{SL}_2(q)$ form a conjugacy class, we obtain that z has a conjugate in $\tilde{H}$ and $\tilde{K}$. If $\lambda \neq 1$, then z has two distinct eigenvalues λ, λ^{-1}. From this it follows that z is $\mathrm{SL}_2(q)$-conjugate to the diagonal matrix $\mathrm{diag}(\lambda, \lambda^{-1}) \in \tilde{K}$. □

The proof of Lemma 3.1 shows that, when $q = 7$, $H = D_8$ and $K = E_q : (q-1)$ are the components of a normal 2-covering of $\mathrm{PSL}_2(7)$. This covering does not appear in the statement of Lemma 3.1, because D_8 is not a maximal subgroup of $\mathrm{PSL}_2(7)$, indeed, $D_8 < S_4$.

Lemma 3.2 *Let q be a prime power. Then $\gamma_w(\mathrm{PSL}_3(q)) = 2$. Moreover, if H and K are the two maximal components of a weak normal 2-covering of $\mathrm{PSL}_3(q)$, then up to $\mathrm{Aut}(\mathrm{PSL}_3(q))$-conjugacy one of the following holds*

(1) $q \neq 4$, $H \cong \left(\frac{q^2+q+1}{\gcd(3,q-1)} \right) : 3$ *is in class $\mathcal{C}_3$ and K is a parabolic subgroup in class $\mathcal{C}_1$,*
(2) $q = 4$, $H \cong \mathrm{SL}_3(2)$ *is in class $\mathcal{C}_5$ and K is a parabolic subgroup in class $\mathcal{C}_1$,*
(3) $q = 4$, $H \cong \mathrm{SL}_3(2)$ *is in class $\mathcal{C}_5$ and $K \cong A_6$ is in class $\mathcal{S}$.*

In (1) the weak normal 2-covering gives rise to two normal coverings; in (2) the weak normal 2-covering gives rise to six normal coverings; in (3) the weak normal 2-covering does not produce a normal covering.

In particular, up to $\mathrm{Aut}(\mathrm{PSL}_3(q))$-conjugacy, there exists a unique weak normal 2-covering of $\mathrm{PSL}_3(q)$ by maximal components when $q \neq 4$; whereas, there are exactly two weak normal 2-coverings of $\mathrm{PSL}_3(q)$ by maximal components when $q = 4$.

Proof We use the list of the maximal subgroups of $\mathrm{PSL}_3(q)$ in [9, Tables 8.3, 8.4] and the notation therein.

When $q = 2$, the proof follows from Lemma 3.1 because $\mathrm{PSL}_3(2) \cong \mathrm{PSL}_2(7)$. When $q = 4$, the proof follows with a computation using the computer algebra system magma [6]. We give some details of our computation. The weak normal 2-covering in (2) splits into six normal 2-coverings of $\mathrm{PSL}_3(q)$. There are three conjugacy classes of subgroups $\mathrm{SL}_3(2)$ and two conjugacy classes of maximal parabolic subgroups and choosing a representative for $\mathrm{SL}_3(2)$ and one for a parabolic gives always rise to a normal covering. Those six normal coverings are distinct up to $\mathrm{PSL}_3(4)$-conjugacy.

For the rest of the proof, we suppose that $q \notin \{2, 4\}$. Let μ be a weak normal covering of $\mathrm{PSL}_3(q)$ of minimum size and with maximal components and $\tilde{\mu}$ be the corresponding weak normal covering of $\mathrm{SL}_3(q)$.

Let x be an element of $\mathrm{SL}_3(q)$ having order $q^2 + q + 1$, that is, a Singer cycle for $\mathrm{SL}_3(q)$. From [9, Tables 8.3, 8.4] and from the fact that $q \neq 4$, we see that the only maximal subgroups of $\mathrm{SL}_3(q)$ containing a Singer cycle form a unique $\mathrm{SL}_3(q)$-conjugacy class of subgroups given by the maximal subgroups in class $\mathcal{C}_3$. Therefore, without loss of generality, we may suppose that $x \in \tilde{H} \cong (q^2+q+1):3$ for some $\tilde{H} \in \tilde{\mu}$ belonging to class $\mathcal{C}_3$.

Let y be an element of $\mathrm{SL}_3(q)$ having order $q^2 - 1$. Using the fact that $q \notin \{2, 4\}$, another case-by-case analysis on the maximal subgroups of $\mathrm{SL}_3(q)$ in [9, Tables 8.2, 8.3] reveals that there are two $\mathrm{SL}_3(q)$-conjugacy classes of maximal subgroups M of $\mathrm{SL}_3(q)$ containing a conjugate of y: namely, the two parabolic subgroups of $\mathrm{SL}_3(q)$, both having structure $E_q^2 : \mathrm{GL}_2(q)$. These two conjugacy classes are fused by a graph automorphism. As $\tilde{H} \not\cong E_q^2 : \mathrm{GL}_2(q)$, the weak normal covering number of $\mathrm{SL}_3(q)$ is 2 and there is a unique possibility for the further component $\tilde{K} \in \tilde{\mu}$.

It is finally easy to see that for every q, $\tilde{H} \cong (q^2 + q + 1) : 3$ and $\tilde{K} \cong E_q^2 : \mathrm{GL}_2(q)$ are components of a normal 2-covering for $\mathrm{SL}_3(q)$ (see [14, Proposition 4.2, Corollary 4.3]) and hence, H and K as in the statement (1) are maximal components of a weak normal 2-covering for $\mathrm{PSL}_3(q)$. That covering is, for $q \neq 4$, the unique weak normal 2-covering for $\mathrm{PSL}_3(q)$ by maximal components. □

The proof of Lemma 3.2 shows that, when $q = 4$, $H \cong 7 : 3$ and K a maximal parabolic subgroup are the components of a weak normal 2-covering of $\mathrm{PSL}_3(4)$. This covering does not appear in the statement of Lemma 3.2 part (1), because $7 : 3$ is not a maximal subgroup of $\mathrm{PSL}_3(4)$, indeed, $7 : 3 < \mathrm{SL}_3(2)$.

Note that $\tilde{G} := \mathrm{SL}_3(3^f)$ admits also the normal covering with components given by $\tilde{H} \cong (q^2 + q + 1) : 3$ and the Levi subgroup L of the parabolic subgroup K [16, Thm. A, Table 1.2]. However, this example is not considered in Table 1.3 because L is not maximal. Of course, this example does appear in our classification of normal 2-coverings in Chap. 11, see Table 11.5.

Lemma 3.3 *Let q be a prime power. Then $\gamma_w(\mathrm{PSL}_4(q)) = 2$. Moreover, if H and K are the two maximal components of a weak normal 2-covering of $\mathrm{PSL}_4(q)$, then up to $\mathrm{Aut}(\mathrm{PSL}_4(q))$-conjugacy $\tilde{H} \cong \mathrm{SL}_2(q^2).(q+1).2$ is in class $\mathcal{C}_3$ and $\tilde{K} \cong E_q^3 : \mathrm{GL}_3(q)$ is in class $\mathcal{C}_1$. This weak normal covering gives rise to two normal coverings.*

Proof We use the list of the maximal subgroups of $\mathrm{PSL}_4(q)$ in [9, Tables 8, 9] and the notation therein. Let μ be a weak normal covering of $\mathrm{PSL}_4(q)$ of minimum size and with maximal components and $\tilde{\mu}$ be the corresponding weak normal covering of $\mathrm{SL}_4(q)$.

Let x be a Singer cycle of $SL_4(q)$. From [9, Tables 8.8, 8.9], we see that there exists a unique $SL_4(q)$-conjugacy class of maximal subgroups containing a Singer cycle, that is, the maximal subgroups in class $\mathcal{C}_3$. Hence there exists $\tilde{H} \in \tilde{\mu}$ belonging to class $\mathcal{C}_3$ and, without loss of generality, we may suppose that

$$x \in \tilde{H} \cong SL_2(q^2).(q+1).2.$$

Let y be an element of $SL_4(q)$ having order $q^3 - 1$. Another case-by-case analysis on the maximal subgroups of $SL_4(q)$ in [9, Tables 8.8, 8.9] reveals that there are two $SL_4(q)$-conjugacy classes of maximal subgroups containing a conjugate of y: namely, the parabolic subgroups of $SL_4(q)$ isomorphic to $E_q^3 : GL_3(q)$. These two conjugacy classes are fused by a graph automorphism. As $\tilde{H} \not\cong E_q^3 : GL_3(q)$, the weak normal covering number of $SL_4(q)$ is 2 and the unique possibility for the further component $\tilde{K} \in \tilde{\mu}$ is given by $\tilde{K} \cong E_q^3 : GL_3(q)$.

The fact that $\tilde{H}$ and $\tilde{K}$ are indeed components of a normal 2-covering for $PSL_4(q)$ follows by [14, Proposition 4.2]. To deduce the number of $\tilde{G}$-classes of normal 2-coverings, we use the observations in Sect. 2.3. Indeed, by [9], the $\text{Aut}(\tilde{G})$-class $\{\tilde{H}^\varphi \mid \varphi \in \text{Aut}(\tilde{G})\}$ is a $\tilde{G}$-conjugacy class, whereas the $\text{Aut}(\tilde{G})$-class $\{\tilde{K}^\varphi \mid \varphi \in \text{Aut}(\tilde{G})\}$ is a union of two distinct $\tilde{G}$-conjugacy classes. Therefore, the weak normal covering $\{\tilde{H}, \tilde{K}\}$ gives rise to two distinct $\tilde{G}$-classes of normal 2-coverings. □

Now, the veracity of Table 1.4 follows from the results in this chapter.

Chapter 4
Unitary Groups

Recall again that given a subgroup X of $\mathrm{PSU}_n(q)$, we denote by $\tilde{X}$ its preimage under the natural projection $\mathrm{SU}_n(q) \to \mathrm{PSU}_n(q)$. Moreover, since the automorphism group of $\mathrm{SU}_n(q)$ projects onto the automorphism group of $\mathrm{PSU}_n(q)$, we may work with the special unitary group $\mathrm{SU}_n(q)$ (see Sect. 2.2).

We start with a lemma that, except for a handful of few small cases, will help us to identify one of the components of a weak normal covering of the unitary group.

Lemma 4.1 ([74, Theorem 1.1]) *Let n be an integer with $n \geq 3$ and $(n, q) \neq (3, 2)$. Let M be a maximal subgroup of $\mathrm{SU}_n(q)$ containing a Singer cycle when n is odd, and a semisimple element of order $q^{n-1} + 1$ and action type $1 \oplus (n - 1)$ when n is even. Then one of the following holds*

(1) *n is odd,*

$$M \cong \mathrm{SU}_{n/k}(q^k) . \frac{q^k + 1}{q + 1} . k,$$

for some prime number k with $k \mid n$, and M is in class $\mathcal{C}_3$,
(2) *n is even and $M \cong \mathrm{GU}_{n-1}(q)$ is in class $\mathcal{C}_1$,*
(3) *$(n, q) = (3, 3)$ and $M \cong \mathrm{PSL}_2(7)$ is in class $\mathcal{S}$,*
(4) *$(n, q) = (3, 5)$ and $M \cong 3.A_7$ is in class $\mathcal{S}$,*
(5) *$(n, q) = (4, 2)$ and $M \cong 3^3.S_4$ is in class $\mathcal{C}_2$,*
(6) *$(n, q) = (4, 3)$ and $M \cong 4.A_7$ is in class $\mathcal{S}$,*
(7) *$(n, q) = (4, 3)$ and $M \cong 4.\mathrm{PSL}_3(4)$ is in class $\mathcal{S}$,*
(8) *$(n, q) = (5, 2)$ and $M \cong \mathrm{PSL}_2(11)$ is in class $\mathcal{S}$,*
(9) *$(n, q) = (6, 2)$ and $M \cong 3.M_{22}$ is in class $\mathcal{S}$.*

4.1 Small Dimensional Unitary Groups

We start our analysis with small dimensional unitary groups $\mathrm{PSU}_n(q)$ with $3 \leq n \leq 6$. Observe that $\mathrm{SU}_n(q)$ is solvable when $(n,q) = (3,2)$ and thus we do not consider that case. For the subgroup structure of $\mathrm{SU}_3(q)$, $\mathrm{SU}_4(q)$, $\mathrm{SU}_5(q)$ and $\mathrm{SU}_6(q)$ we use [9].

Lemma 4.2 *The weak normal covering number of* $\mathrm{PSU}_3(q)$ *for $q > 2$ is at least 2. Moreover, if H and K are maximal components of a weak normal 2-covering of $\mathrm{PSU}_3(q)$, then up to* $\mathrm{Aut}(\mathrm{PSU}_3(q))$*-conjugacy one of the following holds*

(1) $q = 3$, $\tilde{H} \cong \mathrm{PSL}_2(7)$ *is in class $\mathcal{S}$ and* $\tilde{K} \cong \mathrm{GU}_2(q)$ *is in class* $\mathcal{C}_1$,
(2) $q = 3$, $\tilde{H} \cong \mathrm{PSL}_2(7)$ *is in class $\mathcal{S}$ and* $\tilde{K} \cong E_q^{1+2} : (q^2 - 1)$ *is in class* $\mathcal{C}_1$,
(3) q *is a power of 3, $q > 3$,* $\tilde{H} \cong (q^2 - q + 1) : 3$ *is in class $\mathcal{C}_3$ and* $\tilde{K} \cong \mathrm{GU}_2(q)$ *is in class* $\mathcal{C}_1$,
(4) $q = 5$, $\tilde{H} \cong 3.A_7$ *is in class $\mathcal{S}$ and* $\tilde{K} \cong E_q^{1+2} : (q^2 - 1)$ *is in class* $\mathcal{C}_1$,
(5) $q = 5$, $\tilde{H} \cong 3.A_7$ *is in class $\mathcal{S}$ and* $\tilde{K} \cong \mathrm{GU}_2(q)$ *is in class* $\mathcal{C}_1$.

In cases (1)–(3) each weak normal 2-covering of $\mathrm{PSU}_3(q)$ *gives rise to a single normal 2-covering, in case (4) the weak normal 2-covering of* $\mathrm{PSU}_3(q)$ *gives rise to three normal 2-coverings and in case (5) no normal 2-covering arises.*

Proof With the help of the computer algebra system magma [6] we find that for $q \in \{3, 5\}$, the weak normal 2-coverings of $\mathrm{SU}_3(q)$ are the ones reported above. We have also verified that the weak normal coverings in (1), (2) give rise to a single normal 2-covering and that in (4) to three normal 2-coverings, whereas the weak normal covering in (5) does not give rise to a normal covering. See also Sect. 2.3.

Therefore, for the rest of the proof, we suppose $q \notin \{2, 3, 5\}$.

Let μ be a weak normal covering of $\mathrm{PSU}_3(q)$ of minimum size and with maximal components. Let $H \in \mu$ containing a Singer cycle of $\mathrm{PSU}_3(q)$. Then $\tilde{H}$ is a maximal component of a weak normal covering $\tilde{\mu}$ of $\mathrm{SU}_3(q)$ containing a Singer cycle of $\mathrm{SU}_3(q)$. From Lemma 4.1, we have

$$\tilde{H} \cong \left(\frac{q^3 + 1}{q + 1}\right) : 3 = (q^2 - q + 1) : 3.$$

Consider now the elements of $\mathrm{SU}_3(q)$ having order $q^2 - 1$. Using the fact that $q \notin \{2, 3, 5\}$, a case-by-case analysis on the maximal subgroups of $\mathrm{SU}_3(q)$ in [9, Tables 8.5, 8.6] reveals that there are two $\mathrm{SU}_3(q)$-conjugacy classes of maximal subgroups containing elements of order $q^2 - 1$. Namely,

(i) $E_q^{1+2} : (q^2 - 1) \in \mathcal{C}_1$,
(ii) $\mathrm{GU}_2(q) \in \mathcal{C}_1$.

As $\tilde{H}$ is in not in this list, there must exist a further maximal component $\tilde{K}$ of $\tilde{\mu}$ that contains an element of order $q^2 - 1$. Thus $\gamma_w(\mathrm{SU}_3(q)) = \gamma_w(\mathrm{PSU}_3(q)) \geq 2$ and $\tilde{K}$ is in one of the above possibilities (i), (ii).

4.1 Small Dimensional Unitary Groups

Assume now that $\gamma_w(\mathrm{SU}_3(q)) = 2$. Then $\tilde{\mu} = \{\tilde{H}, \tilde{K}\}$. Suppose first that

$$\tilde{K} \cong E_q^{1+2} : (q^2 - 1).$$

We investigate elements having order $q + 1$. Assume, by contradiction, that $\tilde{H}$ contains an element of order $q + 1$. Then, an easy calculation gives

$$q + 1 = \gcd(|\tilde{H}|, q+1) = \gcd(3(q^2 - q + 1), q+1) \mid \gcd(9, q+1) \mid 9.$$

It follows that $q = 8$ and $\tilde{H} \cong 57 : 3$ contains an element of order 9. However a Sylow 3-subgroup of a semidirect product $57 : 3$ cannot be cyclic of order 9. Therefore, we deduce that every element of $\mathrm{SU}_3(q)$ of order $q + 1$ has a conjugate in $\tilde{K}$ via an element in the automorphism group of $\mathrm{SU}_3(q)$. The group $\mathrm{SU}_3(q)$ contains elements g with $\mathrm{o}(g) = q + 1$ and with g having three distinct eigenvalues in $\mathbb{F}_{q^2}$. Indeed, suppose that the Hermitian matrix preserved by $\mathrm{SU}_3(q)$ is

$$J = \begin{pmatrix} 1 & 0 & 0 \\ 0 & 1 & 0 \\ 0 & 0 & 1 \end{pmatrix}.$$

Let $\lambda \in \mathbb{F}_{q^2}$ be an element of order $q + 1$. Then it is easy to verify that

$$\begin{pmatrix} \lambda & 0 & 0 \\ 0 & \lambda^{-1} & 0 \\ 0 & 0 & 1 \end{pmatrix} \in \mathrm{SU}_3(q)$$

and its three eigenvalues are distinct. However, we now show that all elements of $\tilde{K} \cong E_q^{1+2} : (q^2 - 1)$ having order $q + 1$ admit an eigenvalue with multiplicity ≥ 2. Since the outer automorphisms of $\mathrm{SU}_3(q)$ consist of field automorphisms, this property is preserved by conjugation under $\mathrm{Aut}(\mathrm{SU}_3(q))$. Therefore, this case will not arise. Suppose that the Hermitian form preserved by $\mathrm{SU}_3(q)$ has matrix

$$J = \begin{pmatrix} 0 & 0 & 1 \\ 0 & 1 & 0 \\ 1 & 0 & 0 \end{pmatrix}.$$

As $\tilde{K}$ is the stabilizer of a totally singular 1-dimensional subspace of $\mathbb{F}_{q^2}^3$, we may assume that $\tilde{K}$ is the stabilizer of $\langle (1, 0, 0) \rangle_{\mathbb{F}_{q^2}}$. Now, a computation with the matrix J shows that $\tilde{K}$ consists of the matrices

$$\begin{pmatrix} a & 0 & 0 \\ b & c & 0 \\ e & f & g \end{pmatrix},$$

where $a, b, c, e, f, g \in \mathbb{F}_{q^2}$, with $acg = 1$, $g = a^{-q}$, $c^{q+1} = 1$, $cf^q + bg^q = ge^q + f^{q+1} + eg^q = 0$. A Levi complement of $\tilde{K}$ is

$$\left\{ \begin{pmatrix} a & 0 & 0 \\ 0 & a^{q-1} & 0 \\ 0 & 0 & a^{-q} \end{pmatrix} \mid a \in \mathbb{F}_{q^2}^* \right\}.$$

The elements having order $q+1$ in this Levi complement are of the form

$$\begin{pmatrix} a^{q-1} & 0 & 0 \\ 0 & a^{(q-1)^2} & 0 \\ 0 & 0 & a^{-q(q-1)} \end{pmatrix},$$

where a is a generator of the multiplicative group $\mathbb{F}_{q^2}^*$. As

$$a^{-q(q-1)} = a^{-q^2+q} = a^{-1+q} = a^{q-1},$$

we see that the elements of order $q+1$ in the Levi complement have an eigenvalue with multiplicity ≥ 2. Finally, observe that every subgroup generated by an element of order $q+1$ in $\tilde{K}$ is $\tilde{K}$-conjugate to a subgroup of the Levi complement by the Schur–Zassenhaus theorem.

Suppose next that

$$\tilde{K} \cong \mathrm{GU}_2(q).$$

We now consider unipotent elements. The group $\mathrm{SU}_3(q)$ contains unipotent elements u_1 and u_2 with $\dim_{\mathbb{F}_{q^2}} \mathbf{C}_V(u_1) = 1$ and $\dim_{\mathbb{F}_{q^2}} \mathbf{C}_V(u_2) = 2$. The former are precisely the regular unipotent elements. Since every non-identity unipotent element u of $\mathrm{GU}_2(q)$ has the property that $\dim_{\mathbb{F}_{q^2}} \mathbf{C}_V(u) = 2$, we deduce $\tilde{H}$ contains non-identity unipotent elements. Thus,

$$1 \neq \gcd(|\tilde{H}|, q) = \gcd(3, q)$$

and hence q is a power of 3. From [16, Table 1.2], we see that when $q > 3$ is a power of 3 we have that $\tilde{H} \cong (q^2 - q + 1) : 3$ and $\tilde{K} \cong \mathrm{GU}_2(q)$ are indeed the components of a normal 2-covering of $\mathrm{SU}_3(q)$ and hence also of a weak normal 2-covering of $\mathrm{SU}_3(q)$. Section 2.3 justifies the fact that we only have one G-conjugacy class of such coverings. □

Lemma 4.3 *The weak normal covering number of* $\mathrm{PSU}_4(q)$ *is 2. Moreover, if H and K are maximal components of a weak normal 2-covering of* $\mathrm{PSU}_4(q)$*, then up to* $\mathrm{Aut}(\mathrm{PSU}_4(q))$*-conjugacy one of the following holds*

(1) $q = 2$, $\tilde{H} \cong \mathrm{GU}_3(q) \in \mathcal{C}_1$ *and* $\tilde{K} \cong \mathrm{Sp}_4(q) \in \mathcal{C}_5$,
(2) $q = 3$, $\tilde{H} \cong 4.A_7 \in \mathcal{S}$ *and* $\tilde{K} \cong E_q^{1+4} : \mathrm{SU}_2(q) : (q^2 - 1) \in \mathcal{C}_1$,

4.1 Small Dimensional Unitary Groups

(3) $q = 3$, $\tilde{H} \cong \mathrm{GU}_3(q) \in \mathcal{C}_1$ and $\tilde{K} \cong 4.\mathrm{PSU}_4(2) \in \mathcal{S}$,
(4) $q = 3$, $\tilde{H} \cong 4.\mathrm{PSL}_3(4) \in \mathcal{S}$ and $\tilde{K} \cong E_q^{1+4} : \mathrm{SU}_2(q) : (q^2 - 1) \in \mathcal{C}_1$,
(5) $\tilde{H} \cong \mathrm{GU}_3(q)$ and $\tilde{K} \cong E_q^4 : \mathrm{SL}_2(q^2) : (q - 1) \in \mathcal{C}_1$.

Except for (3), *each weak normal 2-covering of* $\mathrm{PSU}_4(q)$ *gives rise to a single normal 2-covering. In case* (3) *no normal 2-covering of* $\mathrm{PSU}_4(q)$ *arises.*

Proof When $q \in \{2, 3\}$, the result follows with a computation with the computer algebra system magma [6]. Therefore, for the rest of the argument, we suppose $q \geq 4$.

Let μ be a weak normal covering of $\mathrm{PSU}_4(q)$ of minimum size and with maximal components. Let $\tilde{H} \in \tilde{\mu}$ containing some element of order $q^3 + 1$ and action type $1 \oplus 3$. From Lemma 4.1, we deduce that

$$\tilde{H} \cong \mathrm{GU}_3(q).$$

Using the fact that $q \notin \{2, 3\}$, a case-by-case analysis on the maximal subgroups of $\mathrm{SU}_4(q)$ in [9, Tables 8.10, 8.11] reveals that there are precisely two $\mathrm{SU}_4(q)$-conjugacy classes of maximal subgroups containing elements of order $(q^4 - 1)/(q + 1)$. Namely,

(i) $E_q^4 : \mathrm{SL}_2(q^2) : (q - 1) \in \mathcal{C}_1$,
(ii) $\mathrm{SL}_2(q^2).(q - 1).2 \in \mathcal{C}_2$.

As $\tilde{H}$ is in not in this list, there must exist a further maximal component $\tilde{K}$ of $\tilde{\mu}$ that contains an element of order $(q^4 - 1)/(q + 1)$. Thus $\gamma_w(\mathrm{SU}_4(q)) = \gamma_w(\mathrm{PSU}_4(q)) \geq 2$. Assume now that $\gamma_w(\mathrm{SU}_4(q)) = 2$. Then $\tilde{\mu} = \{\tilde{H}, \tilde{K}\}$ and $\tilde{K}$ is in one of the above possibilities (i), (ii). We show that necessarily $\tilde{K} \cong E_q^4 : \mathrm{SL}_2(q^2) : (q - 1)$.

Suppose first that q is odd. As neither $\mathrm{GU}_3(q)$ nor $\mathrm{SL}_2(q^2).(q - 1).2$ contain regular unipotent elements, $\tilde{K}$ is isomorphic to $E_q^4 : \mathrm{SL}_2(q^2) : (q - 1)$. By [16, Proposition 5.1],

$$\tilde{H} \cong \mathrm{GU}_3(q) \text{ and } \tilde{K} \cong E_q^4 : \mathrm{SL}_2(q^2) : (q - 1)$$

are indeed the components of a normal 2-covering for $\mathrm{SU}_4(q)$ and hence also of a weak normal 2-covering of $\mathrm{SU}_4(q)$. As usual Sect. 2.3 yields that this weak normal 2-covering gives rise to one G-class of normal 2-coverings. For the rest of our argument we deal with the case q even.

We claim that $\mathrm{SU}_4(q)$ contains an element g with $\mathbf{o}(g) = 2(q^2 - 1)$. To see this, we suppose that the Hermitian form preserved by $\mathrm{SU}_4(q)$ is given by the matrix

$$J := \begin{pmatrix} 0 & 0 & 0 & 1 \\ 0 & 0 & 1 & 0 \\ 0 & 1 & 0 & 0 \\ 1 & 0 & 0 & 0 \end{pmatrix}.$$

Let $\mathbb{F}_{q^2}^* = \langle \lambda \rangle$ and set $\mu := \lambda^q$. Consider now the matrix

$$g := \begin{pmatrix} 0 & 0 & 0 & \lambda \\ 0 & \mu & 0 & 0 \\ 0 & 0 & \mu^{-q} & 0 \\ \lambda^{-q} & 0 & 0 & 0 \end{pmatrix}.$$

An easy computation shows that g preserves the Hermitian form given by J and hence $g \in \mathrm{GU}_4(q)$. Moreover,

$$\det(g) = -\mu^{1-q}\lambda^{1-q} = -\lambda^{q-q^2}\lambda^{1-q} = -\lambda^{1-q^2} = -1 = 1$$

and thus $g \in \mathrm{SU}_4(q)$. Moreover, as

$$g^2 = \begin{pmatrix} \lambda^{1-q} & 0 & 0 & 0 \\ 0 & \mu^2 & 0 & 0 \\ 0 & 0 & \mu^{-2q} & 0 \\ 0 & 0 & 0 & \lambda^{1-q} \end{pmatrix},$$

one sees easily that g has order $2(q^2 - 1)$.

Assume, by contradiction, that g belongs to a stabilizer $\tilde{H} \cong \mathrm{GU}_3(q)$ of some non-degenerate 1-dimensional $\mathbb{F}_{q^2}$-subspace V_1 of V. Then we have $y := g^2 \in \tilde{H}$ and $g \in \mathbf{C}_{\tilde{H}}(y)$. In order to reach a contradiction, it is enough to see that $\mathbf{C}_{\tilde{H}}(y)$ is formed by diagonal matrices. Note that we have $\mathbf{o}(\lambda^{1-q}) = q+1$, $\mathbf{o}(\mu) = q^2 - 1$ and

$$\mathbf{o}(\mu^2) = \mathbf{o}(\mu^{-2q}) = \frac{q^2 - 1}{\gcd(2, q^2 - 1)} = q^2 - 1.$$

Since $q \geq 4$, this implies that λ^{1-q}, μ^2 and μ^{-2q} are distinct, and that the only one among them with order which divides $q + 1$ is λ^{1-q}. Clearly we have $V_{\lambda^{1-q}}(y) = \langle e_1, e_4 \rangle$, $V_{\mu^2}(y) = \langle e_2 \rangle$ and $V_{\mu^{-2q}}(y) = \langle e_3 \rangle$. Moreover, taking into account our Hermitian form, we see that $V = V_{\lambda^{1-q}}(y) \perp (V_{\mu^2}(y) \oplus V_{\mu^{-2q}}(y))$. Since y stabilizes V_1, we have that V_1 is an eigenspace for y corresponding to an eigenvalue whose order must be a divisor of $q + 1$, because V_1 is non-degenerate. Thus that eigenvalue necessarily equals λ^{1-q} and then $V_1 \subset V_{\lambda^{1-q}}(y)$. Pick then $w_1 \in V_{\lambda^{1-q}}(y)$ orthogonal to V_1 and note that the set of vectors $\{w_1, e_2, e_3\}$ is a basis of $V_1^\perp$ consisting of eigenvectors of y with respect to the pairwise distinct eigenvalues λ^{1-q}, μ^2 and μ^{-2q}. Then if

$$d := \begin{pmatrix} \mu^2 & 0 & 0 \\ 0 & \mu^{-2q} & 0 \\ 0 & 0 & \lambda^{1-q} \end{pmatrix},$$

4.1 Small Dimensional Unitary Groups

we have that $\mathbf{C}_{\tilde{H}}(y) \cong \mathbf{C}_{\mathrm{GU}(V_1^\perp)}(d)$ is isomorphic to a subgroup of $\mathbf{C}_{\mathrm{GL}_3(q^2)}(d)$ and since the diagonal entries of d are distinct, $\mathbf{C}_{\mathrm{GL}_3(q^2)}(d)$ is formed only by diagonal matrices. Summing up, we have shown that g does not belong to the stabilizer of a 1-dimensional non-degenerate subspace.

Thus g must be $\mathrm{Aut}(\mathrm{SU}_4(q))$-conjugate to an element of $\tilde{K}$. We now show that $\mathrm{SL}_2(q^2).(q-1).2$ has no element of order $2(q^2-1)$. Suppose that the Hermitian form preserved by $\mathrm{SU}_4(q)$ has matrix

$$J := \begin{pmatrix} 0 & 0 & 1 & 0 \\ 0 & 0 & 0 & 1 \\ 1 & 0 & 0 & 0 \\ 0 & 1 & 0 & 0 \end{pmatrix}.$$

As $\mathrm{SL}_2(q^2).(q-1).2 \in \mathcal{C}_2$, we see that $\mathrm{SL}_2(q^2).(q-1).2$ is the stabilizer of a direct sum decomposition of totally singular subspaces of dimension 2, which we may assume to be $\langle e_1, e_2 \rangle \oplus \langle e_3, e_4 \rangle$, where e_1, e_2, e_3, e_4 is the canonical basis for $V = \mathbb{F}_{q^2}^4$. A direct computation with J shows that the subgroup of $\mathrm{SU}_4(q)$ preserving this direct sum decomposition is

$$\tilde{K} = \left\langle \begin{pmatrix} A & 0 \\ 0 & \bar{A}^{-T} \end{pmatrix}, \begin{pmatrix} 0 & I \\ I & 0 \end{pmatrix} \mid A \in \mathrm{GL}_2(q^2), \det(A) \in \mathbb{F}_q \right\rangle,$$

here we are denoting by $\bar{A}$ the image of A under the involutory field automorphism. We argue by contradiction and we let $x = us = su \in \tilde{K}$ having order $2(q^2-1)$, with u unipotent and s semisimple. If $u \in \mathrm{SL}_2(q^2)$, then $\mathbf{C}_{\mathrm{SL}_2(q^2)}(u)$ is a Sylow 2-subgroup of order q^2 and hence $|\mathbf{C}_{\tilde{K}}(u)|$ divides $q^2(q-1)2$, contradicting the fact that $s \in \mathbf{C}_{\tilde{K}}(u)$ has order q^2-1. Therefore, u is an involution in $\tilde{K} \setminus \mathrm{SL}_2(q^2)$.

By [86, p. 46, Exercise 1], the $\mathrm{SL}_2(q^2)$-conjugacy classes of involutions in $\tilde{K} \setminus \mathrm{SL}_2(q^2)$ are in 1-to-1 correspondence to elements in the non-abelian cohomology set $H^1(\mathrm{Gal}(\mathbb{F}_{q^2}/\mathbb{F}_q), \mathrm{SL}_2(q^2))$. By the Lang–Steinberg theorem $H^1(\mathrm{Gal}(\mathbb{F}_{q^2}/\mathbb{F}_q), \mathrm{SL}_2(q^2))$ consists of a single element (see [86, §2.3, Theorem 1']). Hence we may suppose that

$$u = \begin{pmatrix} 0 & I \\ I & 0 \end{pmatrix}.$$

Now, from a straightforward computation one concludes that

$$\mathbf{C}_{\tilde{K}}(u) = \left\langle \begin{pmatrix} A & 0 \\ 0 & A \end{pmatrix}, \begin{pmatrix} 0 & I \\ I & 0 \end{pmatrix} \mid A \in \mathrm{GL}_2(q^2), \det(A) \in \mathbb{F}_q, A = (\bar{A}^{-1})^T \right\rangle.$$

Now the condition $A = \bar{A}^{-T}$ implies that $A \in \mathrm{GU}_2(q)$. As in dimension 2 we have $\mathrm{GU}_2(q) = (q+1) \times \mathrm{SU}_2(q)$ and as $\det(A) \in \mathbb{F}_q$, we have $A \in \mathrm{SU}_2(q)$. However, there are no elements of order $q^2 - 1$ in $\mathrm{SU}_2(q) \cong \mathrm{SL}_2(q)$. From this argument, one concludes that $\mathrm{SL}_2(q^2).(q-1).2$ does not contain elements of order $2(q^2 - 1)$. Thus we deduce that

$$\tilde{K} \cong E_q^4 : \mathrm{SL}_2(q^2) : (q-1).$$

Finally, by [16, Proposition 5.1],

$$\tilde{H} \cong \mathrm{GU}_3(q) \text{ and } \tilde{K} \cong E_q^4 : \mathrm{SL}_2(q^2) : (q-1)$$

are indeed the components of a normal 2-covering for $\mathrm{SU}_4(q)$ and hence also of a weak normal 2-covering of $\mathrm{SU}_4(q)$. As usual Sect. 2.3 yields that this weak normal 2-covering gives rise to one G-class of normal 2-coverings. □

Lemma 4.4 *The weak normal covering number of* $\mathrm{PSU}_5(q)$ *is at least* 3.

Proof When $q \in \{2, 3\}$, the result follows with a computation with the computer algebra system magma [6]. Therefore, for the rest of the argument, we suppose $q \geq 4$.

We argue by contradiction and suppose that H and K are the maximal components of a weak normal 2-covering μ of $\mathrm{PSU}_5(q)$. Then $\tilde{\mu} = \{\tilde{H}, \tilde{K}\}$ is a weak normal 2-covering with maximal components of $\mathrm{SU}_5(q)$. Since $\mathrm{SU}_5(q)$ contains Singer cycles, from Lemma 4.1, replacing H with K if necessary, we have

$$\tilde{H} \cong \left(\frac{q^5 + 1}{q + 1}\right) : 5.$$

Consider the elements of $\mathrm{SU}_5(q)$ having order $q^4 - 1$. Using the fact that $q \neq 2$, another case-by-case analysis on the maximal subgroups of $\mathrm{SU}_5(q)$ in [9, Tables 8.20, 8.21] reveals that there are two $\mathrm{SU}_5(q)$-conjugacy classes of maximal subgroups containing elements of order $q^4 - 1$. Namely,

(i) $E_q^{4+4} : \mathrm{GL}_2(q^2) \in \mathcal{C}_1$,
(ii) $\mathrm{GU}_4(q) \in \mathcal{C}_1$.

In particular, $\tilde{K}$ must be in one of these two possibilities.

The group $\mathrm{SU}_5(q)$ contains semisimple elements g having order $(q^2 - q + 1) \cdot (q-1)$. To obtain such elements it suffices to start with an orthogonal decomposition $V = W \perp W'$, with $\dim_{\mathbb{F}_{q^2}} W = 3$, $\dim_{\mathbb{F}_{q^2}} W' = 2$, and then take a Singer cycle of order $q^2 - q + 1$ for $\mathrm{SU}_3(q)$ on W (which exists because $q \neq 2$) and a diagonal matrix for $\mathrm{SU}_2(q)$ on W'. Actually, if the Hermitian form induced on W' is

$$\begin{pmatrix} 0 & 1 \\ 1 & 0 \end{pmatrix},$$

then we may suppose that the matrix induced on W' is

$$\begin{pmatrix} a & 0 \\ 0 & a^{-1} \end{pmatrix},$$

for $a \in \mathbb{F}_q^*$ with $\langle a \rangle = \mathbb{F}_q^*$. Observe that, since $q \notin \{2, 3\}$, we have $a \neq a^{-1}$ and thus the only irreducible subspaces of V left invariant by g are W, and two totally singular subspaces of W'. In particular, no $\mathrm{Aut}(\mathrm{SU}_5(q))$-conjugate of g lies in $\mathrm{GU}_4(q)$, because $\mathrm{GU}_4(q)$ is a stabilizer of a non-degenerate 1-dimensional subspace of V.

Finally, since $E_q^{4+4} : \mathrm{GL}_2(q^2)$ contains no elements having order divisible by $q^2 - q + 1$, we see that no $\mathrm{Aut}(\mathrm{SU}_5(q))$-conjugate of g lies in $E_q^{4+4} : \mathrm{GL}_2(q^2)$. □

Lemma 4.5 *The weak normal covering number of* $\mathrm{PSU}_6(q)$ *is at least 2. Moreover, if H and K are maximal components of a weak normal 2-covering of* $\mathrm{PSU}_6(q)$*, then up to* $\mathrm{Aut}(\mathrm{PSU}_6(q))$*-conjugacy one of the following holds*

(1) $q = 2$, $\tilde{H} \cong 3 \times \mathrm{Sp}_6(q) \in \mathcal{C}_5$ *and* $\tilde{K} \cong \mathrm{GU}_5(q) \in \mathcal{C}_1$,
(2) $q = 2$, $\tilde{H} \cong 3.\mathrm{PSU}_4(3) \in \mathcal{S}$ *and* $\tilde{K} \cong \mathrm{GU}_5(q) \in \mathcal{C}_1$.

The weak normal 2-coverings in (1) *and* (2) *do not give rise to normal coverings. In particular* $\gamma(\mathrm{PSU}_6(q)) \geq 3$.

Proof When $q = 2$, the result follows with a computation with the computer algebra system magma [6]. Therefore, for the rest of the argument, we suppose $q \geq 3$.

Let μ be a weak normal covering of $\mathrm{PSU}_6(q)$ of minimum size and with maximal components. Let $\tilde{H} \in \tilde{\mu}$ containing some element of order $q^5 + 1$ and action of type $1 \oplus 5$. From Lemma 4.1, we then have

$$\tilde{H} \cong \mathrm{GU}_5(q).$$

Consider the elements of $\mathrm{SU}_6(q)$ having order $(q^6 - 1)/(q + 1)$. A case-by-case analysis on the maximal subgroups of $\mathrm{SU}_6(q)$ in [9, Tables 8.26, 8.27] reveals that there are three $\mathrm{SU}_6(q)$-conjugacy classes of maximal subgroups containing elements of order $(q^6 - 1)/(q + 1)$. Namely,

(i) $E_q^9 : \mathrm{SL}_3(q^2) : (q - 1) \in \mathcal{C}_1$,
(ii) $\mathrm{SL}_3(q^2).(q - 1).2 \in \mathcal{C}_2$,
(iii) $\mathrm{SU}_2(q^3).(q^2 - q + 1).3 \in \mathcal{C}_3$.

Since none of the above groups is isomorphic to $\tilde{H}$, we deduce that there exists a further component $\tilde{K} \in \tilde{\mu}$. Hence $\gamma_w(\mathrm{PSU}_6(q)) \geq 2$, which confirms the result in [85]. Assume now that there exists a weak normal 2-covering of $\mathrm{PSU}_6(q)$ with maximal components. By the above arguments, we have $\tilde{\mu} = \{\tilde{H}, \tilde{K}\}$ and $\tilde{K}$ must be in one of the three possibilities (i)–(iii).

Let W_1 be an arbitrary 3-dimensional non-degenerate subspace of V and let $W_2 := W_1^\perp$. Now, let $g \in SU_6(q)$ be a semisimple element such that W_1, W_2 are the only g-invariant irreducible subspaces of V. This can be arranged by inducing on W_1 and on W_2 two suitable distinct Singer cycles. Since all elements of $\tilde{H} \cong GU_5(q)$ fix a 1-dimensional non-degenerate subspace of V and since this property is preserved by $\mathrm{Aut}(SU_6(q))$-conjugation, we deduce that g is $\mathrm{Aut}(SU_6(q))$-conjugate to an element in $\tilde{K}$.

As $E_q^9 : SL_3(q^2) : (q-1)$ is the stabilizer of a 3-dimensional totally singular subspace of V, g cannot be $\mathrm{Aut}(SU_6(q))$-conjugate to an element in $E_q^9 : SL_3(q^2) : (q-1)$. Thus $\tilde{K} \not\cong E_q^9 : SL_3(q^2) : (q-1)$.

As $SL_3(q^2).(q-1).2$ is the stabilizer of a direct decomposition $V = V_1 \oplus V_2$ where V_1 and V_2 are 3-dimensional totally singular subspaces of V, g cannot be $\mathrm{Aut}(SU_6(q))$-conjugate to an element in $SL_3(q^2).(q-1).2 \in \mathcal{C}_2$. Thus $\tilde{K} \not\cong SL_3(q^2).(q-1).2$.

Therefore $\tilde{K}$ is isomorphic to the extension field subgroup $SU_2(q^3).(q^2-q+1).3$. Now, for every $i \in \{1, 2, 3, 4, 5\}$, $SU_6(q)$ contains a unipotent element u with $\dim_{\mathbb{F}_{q^2}} \mathbf{C}_V(u) = i$. Clearly, every unipotent element u of $\tilde{H} \cong GU_5(q)$ satisfies $\dim_{\mathbb{F}_{q^2}} \mathbf{C}_V(u) \geq 2$. Therefore, all unipotent elements u of $SU_6(q)$ satisfying $\dim_{\mathbb{F}_{q^2}} \mathbf{C}_V(u) = 1$ are conjugate, via an element in $\mathrm{Aut}(SU_6(q))$, to an element of $\tilde{K}$. As $\tilde{K} \cong SU_2(q^3).(q^2-q+1).3$ and as every unipotent element $u \in SU_2(q^3)$ satisfies $\dim_{\mathbb{F}_{q^2}} \mathbf{C}_V(u) \geq 2$, this case can arise only when $\gcd(q, (q^2-q+1)3) \neq 1$, that is, when q is a power of 3. Summing up, q is a power of 3 and $\tilde{K} \cong SU_2(q^3).(q^2-q+1).3$. Without loss of generality, we may suppose that the Hermitian form left invariant by $SU_6(q)$ is given via the matrix

$$\begin{pmatrix} 0 & 1 & 0 & 0 & 0 & 0 \\ 1 & 0 & 0 & 0 & 0 & 0 \\ 0 & 0 & 0 & 0 & 1 & 0 \\ 0 & 0 & 0 & 0 & 0 & 1 \\ 0 & 0 & 1 & 0 & 0 & 0 \\ 0 & 0 & 0 & 1 & 0 & 0 \end{pmatrix}.$$

Using this Hermitian form, it is readily seen that $SU_6(q)$ contains the matrix

$$z := \begin{pmatrix} 1 & 1 & 0 & 0 \\ 0 & 1 & 0 & 0 \\ 0 & 0 & s & 0 \\ 0 & 0 & 0 & (\bar{s}^{-1})^T \end{pmatrix},$$

where $s \in SL_2(q^2)$ is a Singer cycle having order $q^2 + 1$. The only 1-dimensional subspace of V stabilized by z is $\langle (0, 1, 0, 0, 0, 0) \rangle$, which is totally singular. Therefore z has no conjugate in $\tilde{H} \cong GU_5(q)$, via elements in $\mathrm{Aut}(SU_6(q))$. By construction, $\mathbf{o}(z) = 3(q^2+1)$. With a computation, using Lemma 2.2, we obtain

$$\gcd(q^2+1, |\tilde{K}|) = \gcd(q^2+1, q^3(q^6-1)(q^2-q+1)3)$$
$$= \gcd(q^2+1, (q^6-1)(q^2-q+1))$$
$$= \gcd(2, q-1) = 2.$$

Therefore, z is not $\mathrm{Aut}(\mathrm{SU}_6(q))$-conjugate to an element of $\tilde{K}$. □

4.2 Large Dimensional Unitary Groups

In this section we deal with large dimensional unitary groups $\mathrm{SU}_n(q)$ with $n \geq 7$.

Let t be a Bertrand number for n. As $n \geq 7$, t is an odd prime with $\gcd(t, n) = 1$, $t \geq 5$ and $n/2 < t \leq n-2$. Recall that if $n \geq 8$, then the stronger inequality $n/2 < t \leq n-3$ holds.

Moreover, from Lemma 2.2 (2) and (3), we have $\gcd(q^t+1, q^{n-t}+(-1)^n) = q+1$. In order to find the components of a weak normal covering of $\mathrm{SU}_n(q)$, we consider a Bertrand element $z \in \mathrm{SU}_n(q)$ such that

- $\mathbf{o}(z) = \frac{(q^t+1)(q^{n-t}+(-1)^n)}{\gcd(q^t+1, q^{n-t}+(-1)^n)} = \frac{(q^t+1)(q^{n-t}+(-1)^n)}{q+1}$ (see Table 2.2),
- the action of z on V is of type $t \oplus \frac{n-t}{2} \oplus \frac{n-t}{2}$ if n is odd and of type $t \oplus (n-t)$ if n is even. We write $V = V_t \perp W_1 \perp W_2$ when n is odd and $V = V_t \perp W$ when n is even, where $\dim_{\mathbb{F}_{q^2}} V_t = t$, $\dim_{\mathbb{F}_{q^2}} W_1 = \dim_{\mathbb{F}_{q^2}} W_2 = (n-t)/2$, $\dim_{\mathbb{F}_{q^2}} W = n-t$ and V_t, W_1, W_2, W are z-invariant submodules of V, with W and V_t non-degenerate and W_1, W_2 totally singular.
- z induces a matrix of order q^t+1 on V_t, of order $q^{n-t}+1$ on W and of order $q^{n-t}-1$ on both W_1 and W_2.

We first justify the existence of the Bertrand elements.

Let first n be even. Then both t and $n-t$ are odd. Pick a generator λ for the subgroup of order $q+1$ in $\mathbb{F}_{q^2}^*$. Then, by Proposition 2.5 (3), there exist Singer cycles $s_1 \in \mathrm{GU}_t(q)$ and $s_2 \in \mathrm{GU}_{n-t}(q)$ such that $\det(s_1) = \lambda$ and $\det(s_2) = \lambda^{-1}$. Thus, by the embedding of $\mathrm{GU}_t(q) \perp \mathrm{GU}_{n-t}(q)$ in $\mathrm{GU}_n(q)$, we get an element $z \in \mathrm{SU}_n(q)$ having all the properties required to be a Bertrand element.

Let next n be odd. The embedding of $\mathrm{GL}_{\frac{n-t}{2}}(q^2)$ in $\mathrm{GU}_{n-t}(q)$ determines a block matrix x of order $q^{n-t}-1$ and determinant $b = a^{-q+1}$, where $\langle a \rangle = \mathbb{F}_{q^2}^*$. Since b generates the subgroup of order $q+1$ of $\mathbb{F}_{q^2}^*$, by Proposition 2.5 (3), there exists a Singer cycle s of $\mathrm{GU}_t(q)$ having determinant equal to b^{-1}. Thus the block diagonal matrix constructed using x and s gives the required Bertrand element z.

In the next result, we use [66, Section 4.1] for the labeling of the parabolic subgroups of $\mathrm{SU}_n(q)$.

Lemma 4.6 *Let M be a maximal subgroup of $\mathrm{SU}_n(q)$ with $n \geq 7$ containing a Bertrand element z as in Table 2.2 and satisfying the conditions above. Then one of the following holds*

(1) *M is of type $\mathrm{GU}_t(q) \perp \mathrm{GU}_{n-t}(q)$ and lies in class $\mathcal{C}_1$,*
(2) *n is odd, M is a parabolic subgroup of type $P_{(n-t)/2}$ and lies in class $\mathcal{C}_1$.*

Proof As $t \geq 5$, from Zsigmondy's theorem and from Lemma 2.3, we deduce that

$$\varnothing \neq P_{2t}(q) \subseteq P_t(q^2). \tag{4.1}$$

Let $r \in P_{2t}(q)$. Then $r \mid q^t + 1$ and $r \nmid q + 1$. Hence $r \mid \frac{q^t+1}{q+1}$ and thus $r \mid \mathbf{o}(z) = \frac{q^t+1}{q+1}(q^{n-t} + (-1)^n)$. Moreover, by (2.4), we have $r \geq 2t + 1$. In particular, z is a $ppd(n, q^2; t)$-element with order divisible by an $r \in P_t(q^2)$ with $r \geq 2t + 1$. It follows that M is a $ppd(n, q^2; t)$-group and, by Theorem 2.6, we search M among the groups in the Examples 2.1–2.9 of [53], with $n \geq 7$ and $r \geq 2t + 1$. From the definition of Bertrand number and from $n \geq 7$, we have either $n/2 < t \leq n - 3$ or $(n, t) = (7, 5)$. In any case we also have $\gcd(t, n) = 1$. Now, looking at Tables 2–8 of [53], it is easily checked that $r \neq 2t + 1$ since $t \leq n - 2$. These facts rule out the groups of Examples 2.3, 2.4, 2.5, 2.6, 2.7, 2.8, 2.9, because they are all given under at least one of the conditions:

- $t > n - 3$ and $n \neq 7$, or
- $t = n - 3$ even, or
- $r = t + 1$, or
- $r = 2t + 1$.

Hence we may reduce our considerations to Examples 2.1–2.2, that is, to $M \in \mathcal{C}_1$ or $M \in \mathcal{C}_5$.

The Subgroup M Lies in Class $\mathcal{C}_1$ We have two cases to consider: M is the stabilizer of a proper totally singular subspace or of a proper non-degenerate subspace of V. We consider first the second possibility. Thus M is of type $\mathrm{GU}_m(q) \perp \mathrm{GU}_{n-m}(q)$, with $1 \leq m < n/2$. Recall that the action of z on V is of type $t \oplus \frac{n-t}{2} \oplus \frac{n-t}{2}$ or $t \oplus n-t$ depending on whether n is odd or even. When n is even, z leaves invariant exactly two proper subspaces of V, one having dimension t and the other having dimension $n-t$; moreover, these subspaces are both non-degenerate. Therefore, if $z \in M$, then $t = n - m$, that is, $m = n - t$ and we obtain part (1). Suppose n odd. By definition, V has exactly three irreducible $\mathbb{F}_{q^2}\langle z \rangle$-submodules: namely, W_1, W_2 and V_t, where V_t and $W_1 \oplus W_2$ are non-degenerate and W_1, W_2 are totally singular. Therefore, if $z \in M$, then $t = n - m$, that is, $m = n - t$ and we obtain part (1). We consider now the case that M is the stabilizer of a proper totally singular subspace of V. Thus M is a parabolic subgroup P_m, with $1 \leq m \leq \lfloor n/2 \rfloor$. When n is even, the two proper subspaces of V left invariant by z are both non-degenerate and hence z lies in no parabolic subgroup. Thus n is odd. The only totally singular proper subspaces of V

4.2 Large Dimensional Unitary Groups

which are z-invariant are W_1 and W_2. As $\dim_{\mathbb{F}_{q^2}} W_1 = \dim_{\mathbb{F}_{q^2}} W_2 = (n-t)/2$, we deduce $m = (n-t)/2$ and we obtain part (2).

The Subgroup M Lies in Class $\mathcal{C}_5$ By Lemma 2.7, we have $M \leq \mathrm{GL}_n(q^{\delta/k})$, where $\delta = 2$ and k is a prime with $k \mid 2f$. When $k = 2$, the possibility $M \in \mathcal{C}_5$ is ruled out by (4.1) and Lemma 2.7 (2). When $k \geq 3$, as $kt \geq 15$, Lemma 2.7 (2) applies because $P_{kt}(q^{2/k}) \neq \emptyset$. □

Proposition 4.7 *For every $n \geq 7$, the weak normal covering number of $\mathrm{PSU}_n(q)$ is at least 3.*

Proof As usual, we argue with $\mathrm{SU}_n(q)$. Let $\tilde{H}$ be a component of a weak normal covering of $\mathrm{SU}_n(q)$ containing a semisimple element having order $(q^n + 1)/(q + 1)$ and of type n in its action on V when n is odd, and having order $q^{n-1} + 1$ and of type $1 \oplus (n-1)$ in its action on V when n is even. Moreover, let $\tilde{K}$ be a component containing the Bertrand element defined in Lemma 4.6.

Assume first n even. From Lemma 4.1, we have $\tilde{H} \cong \mathrm{GU}_{n-1}(q)$ and, from Lemma 4.6, we have that $\tilde{K}$ is of type $\mathrm{GU}_t(q) \perp \mathrm{GU}_{n-t}(q)$. Let $x \in \mathrm{SU}_n(q)$ be a semisimple element having order $(q^n - 1)/(q + 1)$ and having type $\frac{n}{2} \oplus \frac{n}{2}$ on V. Thus $V = W_1 \perp W_2$, $\dim_{\mathbb{F}_{q^2}} W_1 = \dim_{\mathbb{F}_{q^2}} W_2 = n/2$ and W_1, W_2 are x-invariant. We also choose x so that the matrix induced in its action on W_1 and on W_2 has order $(q^n - 1)/(q + 1)$. Now, the only proper x-invariant subspaces of V are W_1 and W_2 and hence x cannot have an $\mathrm{Aut}(\mathrm{SU}_n(q))$-conjugate in $\tilde{H} \cong \mathrm{GU}_{n-1}(q)$ because $n/2 \notin \{1, n-1\}$. Similarly, x cannot have an $\mathrm{Aut}(\mathrm{SU}_n(q))$-conjugate in $\tilde{K}$ because $t \neq n/2$. Therefore a weak normal covering of $\mathrm{SU}_n(q)$ contains at least one more component.

Assume n odd. From Lemma 4.1, we have $\tilde{H} \cong \mathrm{SU}_{n/k}(q^k).((q^k + 1)/(q + 1)).k$ for some prime divisor k of n and, from Lemma 4.6, we have that $\tilde{K}$ is a parabolic subgroup $P_{(n-t)/2}$ or is of type $\mathrm{GU}_t(q) \perp \mathrm{GU}_{n-t}(q)$. Let $x \in \mathrm{SU}_n(q)$ be a semisimple element having order $q^{n-2}+1$ and having type $1 \oplus 1 \oplus (n-2)$ on V. Thus $V = W_1 \perp W_2 \perp W$, $\dim_{\mathbb{F}_{q^2}} W_1 = \dim_{\mathbb{F}_{q^2}} W_2 = 1$, $\dim_{\mathbb{F}_{q^2}} W = n-2$ and W_1, W_2, W are x-invariant. We also choose x so that the matrix induced in its action on W_1 and on W_2 has order $q+1$ and the matrix induced by x on W has order $q^{n-2}+1$. Now, the only proper x-invariant subspaces of V are W_1, W_2 and W. Hence x cannot have an $\mathrm{Aut}(\mathrm{SU}_n(q))$-conjugate in $\tilde{K}$ because x does not stabilize a totally isotropic subspace of dimension $(n-t)/2$, nor a non-degenerate subspace of dimension t. Suppose x is $\mathrm{Aut}(\mathrm{SU}_n(q))$-conjugate to an element in $\tilde{H}$. Then $q^{n-2} + 1$ divides the order of $\tilde{H}$. From Zsigmondy's theorem, we may choose $q_{2(n-2)} \in P_{2(n-2)}(q)$. Now, the definition of primitive prime divisor yields that $q_{2(n-2)}$ is relatively prime to the order of $\mathrm{SU}_{n/k}(q^k).((q^k + 1)/(q + 1))$. Thus $q_{2(n-2)}$ divides k and hence $q_{2(n-2)} \leq k$. However, (2.4) yields $q_{2(n-2)} \geq 2(n-2) + 1 = 2n - 3$ and hence $k \geq 2n - 3 > n$, a contradiction. Therefore a weak normal covering of $\mathrm{SU}_n(q)$ contains at least one more component. Thus $\gamma_w(\mathrm{SU}_n(q)) = \gamma_w(\mathrm{PSU}_n(q)) \geq 3$. □

Now, the veracity of Table 1.5 follows from the results in this chapter.

Chapter 5
Symplectic Groups

As usual, given a subgroup X of $\mathrm{PSp}_n(q)$, we denote by $\tilde{X}$ its preimage under the natural projection $\mathrm{Sp}_n(q) \to \mathrm{PSp}_n(q)$. Exactly as for the unitary groups, since the automorphism group $\mathrm{Aut}(\mathrm{Sp}_n(q))$ projects onto the automorphism group of $\mathrm{PSp}_n(q)$, we may work with $\mathrm{Sp}_n(q)$. Extra care must be taken when $n = 4$ and q is even, because graph-field automorphisms of $\mathrm{Sp}_4(q)$ do not act on the vector space V.

Lemma 5.1 ([74, Theorem 1.1]) *Let n be an integer with $n \geq 4$, $(n, q) \neq (4, 2)$ and let M be a maximal subgroup of $\mathrm{Sp}_n(q)$ containing a Singer cycle. Then one of the following holds*

(1) $M \cong \mathrm{Sp}_{n/k}(q^k).k$ *for some prime number k with $k \mid n$ and n/k even, and $M \in \mathcal{C}_3$,*
(2) $nq/2$ *is odd and* $M \cong \mathrm{GU}_{n/2}(q).2$ *is in class $\mathcal{C}_3$,*
(3) q *is even and* $M \cong \mathrm{SO}_n^-(q)$ *is in class $\mathcal{C}_8$,*
(4) $(n, q) = (8, 2)$ *and* $M \cong \mathrm{PSL}_2(17)$ *is in class $\mathcal{S}$,*
(5) $(n, q) = (4, 3)$ *and* $M \cong 2_-^{1+4}.A_5$ *is in class $\mathcal{C}_6$.*

5.1 Small Dimensional Symplectic Groups

We start our analysis with small dimensional symplectic groups $\mathrm{PSp}_n(q)$ with $4 \leq n \leq 8$. Observe that $\mathrm{Sp}_4(2) \cong S_6$ is not simple and that $\mathrm{PSp}_4(3) \cong \mathrm{PSU}_4(2)$. For the subgroup structure of $\mathrm{Sp}_4(q)$, $\mathrm{Sp}_6(q)$ and $\mathrm{Sp}_8(q)$ we use [9].

Lemma 5.2 *Let q be odd. The weak normal covering number of $\mathrm{PSp}_4(q)$ is at least 2. Moreover, if H and K are maximal components of a weak normal 2-covering* of $\mathrm{PSp}_4(q)$, then we have $q = 3$ and up to $\mathrm{Aut}(\mathrm{PSp}_4(q))$-conjugacy one of the following holds

(1) $\tilde{H} \cong E_3^{1+2} : (2 \times \mathrm{Sp}_2(3)) \in \mathcal{C}_1$ and $\tilde{K} \cong \mathrm{Sp}_2(9) : 2 \in \mathcal{C}_3$,
(2) $\tilde{H} \cong E_3^{1+2} : (2 \times \mathrm{Sp}_2(3)) \in \mathcal{C}_1$ and $\tilde{K} \cong 2_-^{1+4}.A_5 \in \mathcal{C}_6$.

Each of the two weak normal 2-coverings of $\mathrm{PSp}_4(3)$ *in* (1) *and* (2) *gives rise to a single normal 2-covering.*

Proof It follows with the help of the computer algebra system magma [6] that $\mathrm{Sp}_4(3)$ does admit weak normal 2-coverings and the only weak normal 2-coverings are reported above. These are also normal coverings. Therefore, for the rest of the proof, we suppose $q \neq 3$.

Let μ be a weak normal 2-covering of $\mathrm{PSp}_4(q)$ with maximal components. Let $\tilde{H}$ be a component of $\tilde{\mu}$ containing a Singer cycle of $\mathrm{Sp}_4(q)$. From Lemma 5.1, we have

$$\tilde{H} \cong \mathrm{Sp}_2(q^2) : 2.$$

Observe that, from Lemma 5.1, $\tilde{H} \in \mathcal{C}_3$ and hence $\tilde{H} = X \rtimes Y$, where $X \cong \mathrm{Sp}_2(q^2)$, $|Y| = 2$ and the non-identity element of Y acts on X via an involutory field automorphism.

Write $q = p^f$, with p an odd prime number and $f \geq 1$. Using the fact that $q \geq 5$, a case-by-case analysis on the maximal subgroups of $\mathrm{Sp}_4(q)$ in [9, Tables 8.12, 8.13] reveals that there are four $\mathrm{Sp}_4(q)$-conjugacy classes of maximal subgroups containing elements of order $p(q + 1)$. Namely,

(i) $E_q^{1+2} : ((q - 1) \times \mathrm{Sp}_2(q)) \in \mathcal{C}_1$,
(ii) $E_q^3 : \mathrm{GL}_2(q) \in \mathcal{C}_1$,
(iii) $\mathrm{Sp}_2(q)^2 : 2 \in \mathcal{C}_2$,
(iv) $\mathrm{GU}_2(q).2 \in \mathcal{C}_3$.

As $\tilde{H}$ is in not in this list, the weak normal covering number of $\mathrm{Sp}_4(q)$ is at least 2. Assume now, by contradiction, that $\gamma_w(\mathrm{PSp}_4(q)) = 2$. Then we have $\tilde{\mu} = \{\tilde{H}, \tilde{K}\}$, with $\tilde{K}$ containing an element of order $p(q + 1)$. In particular, $\tilde{K}$ must be in one of the above four possibilities (i)–(iv).

We now consider elements of order $p(q - 1)$ in $\mathrm{Sp}_4(q)$. Recall that $\tilde{H} \cong \mathrm{Sp}_2(q^2) : 2$ and that $\mathrm{Sp}_2(q^2) \cong \mathrm{SL}_2(q^2)$. A non-identity unipotent element u of $\mathrm{SL}_2(q^2)$ has centralizer of order $2q^2$, given by the product of the Sylow p-subgroup of $\mathrm{SL}_2(q^2)$ containing u and of $Z(\mathrm{SL}_2(q^2)) \cong C_2$.

From this we easily deduce that $\tilde{H}$ contains no elements of order $p(q - 1)$. Indeed assume that there exist $y \in \tilde{H}$ with $o(y) = p(q - 1)$. Then $u := y^{q-1}$ is a non-identity unipotent element of order p. Since p is odd, we necessarily have that $u \in \mathrm{SL}_2(q^2)$. By the above argument, we get $|\mathbf{C}_{\mathrm{SL}_2(q^2)}(u)| = 2q^2$ and thus $|\mathbf{C}_{\tilde{H}}(u)|$ divides $4q^2$. On the other hand, we have $\langle y \rangle \subseteq \mathbf{C}_{\tilde{H}}(u)$ and thus $p(q - 1) \mid 4q^2$. It

5.1 Small Dimensional Symplectic Groups

follows that $q-1 \mid 4$ and so $q \le 5$. On the other hand it is directly checked[1] that $SL_2(25):2$ does not contain an element of order $5 \cdot 4 = 20$.

Therefore, elements of order $p(q-1)$ all have an $\mathrm{Aut}(Sp_4(q))$-conjugate in $\tilde{K}$. We show now that $GU_2(q).2$ admits no element of order $p(q-1)$. Let $GU_2(q)$ be defined by the Hermitian form having matrix

$$J := \begin{pmatrix} 0 & 1 \\ 1 & 0 \end{pmatrix}$$

and note that a non-identity unipotent element of order p of $GU_2(q)$ is conjugate to $u = \begin{pmatrix} 1 & a \\ 0 & 1 \end{pmatrix}$, for some $a \in \mathbb{F}_{q^2}^*$ with $a^q + a = 0$. By an elementary computation one checks that, if $\mathbb{F}_{q^2}^* = \langle t \rangle$, then

$$C_{GU_2(q)}(u) \le \left\{ \begin{pmatrix} \alpha & \beta \\ 0 & \alpha \end{pmatrix} \mid \alpha \in \langle t^{q-1} \rangle, \beta \in \mathbb{F}_{q^2} \right\} \le GL_2(q^2),$$

and thus

$$|C_{GU_2(q)}(u)| \mid (q+1)q^2. \tag{5.1}$$

Assume now, by contradiction, that there exists $x \in GU_2(q).2$ with $o(x) = p(q-1)$. Then $x^2 \in GU_2(q)$ has order $p(q-1)/2$ and thus there exists a non-identity unipotent element u of order p whose centralizer in $GU_2(q)$ has size divisible by $p(q-1)/2$. By (5.1), we deduce $p(q-1)/2 \mid (q+1)q^2$ and therefore $(q-1)/2 \mid (q+1)$. It follows that $(q-1)/2 \mid \gcd(q-1, q+1) = 2$ and thus $q = 5$. On the other hand it is directly checked[2] that $GU_2(5).2$ does not contain an element of order $5 \cdot 4 = 20$.

Since $GU_2(q).2$ has no elements of order $p(q-1)$, $\tilde{K}$ cannot be isomorphic to $GU_2(q).2$ and hence we exclude this case from any further analysis.

Now, $Sp_4(q)$ admits two types of elements having order $p(q+1)$: elements admitting some eigenvalues in the ground field $\mathbb{F}_q$ and elements admitting no eigenvalues in the ground field $\mathbb{F}_q$. This can be more easily seen by considering two symplectic forms. Using the symplectic form having matrix

$$\begin{pmatrix} 0 & 1 & 0 & 0 \\ -1 & 0 & 0 & 0 \\ 0 & 0 & 0 & 1 \\ 0 & 0 & -1 & 0 \end{pmatrix},$$

---

[1] Recall that $SL_2(25):2$ is obtained from $SL_2(25)$ extending by the action of the Galois group of the field $\mathbb{F}_{25}$.
[2] This can be checked with a computer computation by constructing $GU_2(5).2 \in \mathcal{C}_3$ in $Sp_4(5)$.

we can check with an easy computation that the matrices

$$\begin{pmatrix} 1 & 1 & 0 \\ 0 & 1 & 0 \\ 0 & 0 & s \end{pmatrix},$$

where $s \in \mathrm{SL}_2(q)$ is a Singer cycle, are symplectic of order $p(q+1)$ and have some eigenvalues in the ground field $\mathbb{F}_q$. Whereas, using the symplectic form having matrix

$$\begin{pmatrix} 0 & 0 & 1 & 0 \\ 0 & 0 & 0 & 1 \\ -1 & 0 & 0 & 0 \\ 0 & -1 & 0 & 0 \end{pmatrix},$$

we can check with a computation that the 2×2 block matrices

$$\begin{pmatrix} s & s \\ 0 & s \end{pmatrix},$$

where $s \in \mathrm{O}_2^-(q)$ is a Singer cycle are symplectic of order $p(q+1)$ and have no eigenvalues in the ground field $\mathbb{F}_q$.

However, the elements having order $p(q+1)$ in $\mathrm{Sp}_2(q)^2 : 2$ and in $E_q^{1+2} : ((q-1) \times \mathrm{Sp}_2(q))$ are only of the first type, that is, admit an eigenvalue in $\mathbb{F}_q$. On the other hand, the elements having order $p(q+1)$ in $E_q^3 : \mathrm{GL}_2(q)$ are only of the second type, that is, admit no eigenvalues in $\mathbb{F}_q$. Since having eigenvalues in $\mathbb{F}_q$ is a property preserved by $\mathrm{Aut}(\mathrm{Sp}_4(q))$-conjugacy, we deduce that whatever one chooses for $\tilde{K}$, in the possibilities (i), (ii), (iii) and (iv) above, the $\mathrm{Aut}(\mathrm{Sp}_4(q))$-conjugates of $\tilde{K}$ contain only one type of elements of order $p(q+1)$, and the contradiction is found. □

We now deal with even characteristic 4-dimensional symplectic groups. Here the result is rather different from Lemma 5.2 because of the peculiar subgroup lattice arising in characteristic 2. Recall that $\mathrm{PSp}_4(q) = \mathrm{Sp}_4(q)$, when q is even. We start with a technical lemma which follows from the work of Enomoto [35] on the character table of $\mathrm{Sp}_4(q)$. However, to explain how the tables in [35] can be used to deduce this lemma would take us too far astray. Therefore, here, we give a direct proof.

Lemma 5.3 *Let f be a positive integer and let $q := 2^f$. The following holds*

(1) $\mathrm{Sp}_4(q)$ *contains a family* $\mathcal{D} \neq \emptyset$ *of elements h having order $q+1$ and with* $|\mathbf{C}_{\mathrm{Sp}_4(q)}(h)| = (q+1)^2$ *if and only if $f > 1$,*
(2) *for every $h \in \mathcal{D}$, $|\mathbf{N}_{\mathrm{Sp}_4(q)}(\langle h \rangle) : \mathbf{C}_{\mathrm{Sp}_4(q)}(h)| \in \{2, 4\}$,*

5.1 Small Dimensional Symplectic Groups

(3) when $f > 2$, $\mathrm{Sp}_4(q)$ contains elements h_1 and h_2 belonging to $\mathcal{D}$ with $|\mathbf{N}_{\mathrm{Sp}_4(q)}(\langle h_1 \rangle) : \mathbf{C}_{\mathrm{Sp}_4(q)}(h_1)| = 2$ and $|\mathbf{N}_{\mathrm{Sp}_4(q)}(\langle h_2 \rangle) : \mathbf{C}_{\mathrm{Sp}_4(q)}(h_2)| = 4$,

(4) when f is even, each element $h \in \mathcal{D}$ which is $\mathrm{Aut}(\mathrm{Sp}_4(q))$-conjugate to an element of the maximal subgroup $\mathrm{Sp}_4(\sqrt{q}) \in \mathcal{C}_5$ satisfies $|\mathbf{N}_{\mathrm{Sp}_4(q)}(\langle h \rangle) : \mathbf{C}_{\mathrm{Sp}_4(q)}(h)| = 4$.

Proof We let the symplectic form of $\mathrm{Sp}_4(q)$ be defined by the matrix

$$\begin{pmatrix} 0 & 1 & 0 & 0 \\ 1 & 0 & 0 & 0 \\ 0 & 0 & 0 & 1 \\ 0 & 0 & 1 & 0 \end{pmatrix}.$$

We first claim that

(†) The centralizer of every element $h \in \mathrm{Sp}_4(q)$ of order $q+1$ has size divisible by $(q+1)^2$.

(††) Moreover, if the size of the centralizer of h is exactly $(q+1)^2$, then h is conjugate to a matrix of the form

$$h_{\ell_1,\ell_2} = \begin{pmatrix} \zeta^{\ell_1} & 0 \\ 0 & \zeta^{\ell_2} \end{pmatrix}, \qquad (5.2)$$

where $\zeta \in \mathrm{Sp}_2(q)$ is a Singer cycle, $\ell_1, \ell_2 \in \{1, \ldots, q\}$ and $\mathbf{o}(h_{\ell_1,\ell_2}) = \mathrm{lcm}\{\mathbf{o}(\zeta^{\ell_1}), \mathbf{o}(\zeta^{\ell_2})\} = q+1$.

Let $h \in \mathrm{Sp}_4(q)$ be of order $q+1$. Then h is semisimple and hence V decomposes as a direct sum of irreducible $\mathbb{F}_q \langle h \rangle$-modules. Thus we have four cases to consider:

- V is irreducible,
- $V = V_1 \oplus V_2 \oplus V_3 \oplus V_4$ with $\dim_{\mathbb{F}_q} V_i = 1$, for $i \in \{1, \ldots, 4\}$,
- $V = V_1 \oplus V_2 \oplus V_3$ with $\dim_{\mathbb{F}_q} V_1 = \dim_{\mathbb{F}_q} V_2 = 1$ and $\dim_{\mathbb{F}_q} V_3 = 2$,
- $V = V_1 \oplus V_2$ with $\dim_{\mathbb{F}_q} V_1 = \dim_{\mathbb{F}_q} V_2 = 2$.

The first case does not arise because in that case we would have $\mathbf{o}(h) \mid q^2+1$ and hence $q+1 = \mathbf{o}(h) \mid \gcd(q^2+1, q+1) = 1$, which is a contradiction. Similarly the second case does not arise because in that case we would have $\mathbf{o}(h) \mid q-1$ and hence $q+1 = \mathbf{o}(h) \mid \gcd(q-1, q+1) = 1$, which is a contradiction. In the third case, by Sect. 2.5, we have that V_3 is non-degenerate, $V_1 \oplus V_2 = V_3^\perp$ and V_1, V_2 are totally isotropic. Therefore the centralizer of h has order $(q+1)|\mathrm{Sp}_2(q)|$, which is a proper multiple of $(q+1)^2$. Note that the factor $q+1$ arises from the centralizer of the Singer cycle induced on V_3, and $\mathrm{Sp}_2(q)$ arises from the centralizer of the matrix induced on $V_1 \oplus V_2$. Finally, in the fourth case h is $\mathrm{Sp}_4(q)$-conjugate to a 2×2-block matrix of the form

$$\begin{pmatrix} \xi & 0 \\ 0 & \mu \end{pmatrix},$$

where $\xi, \mu \in \mathrm{Sp}_2(q)$ have irreducible characteristic polynomials and

$$q + 1 = \mathrm{lcm}\{\mathbf{o}(\xi), \mathbf{o}(\mu)\}.$$

Thus ξ and μ are powers of suitable Singer cycles. Let $\zeta \in \mathrm{Sp}_2(q)$ be a fixed Singer cycle in $\mathrm{Sp}_2(q)$. Since all Singer cycles in $\mathrm{Sp}_2(q)$ are $\mathrm{Sp}_2(q)$-conjugate to some Singer cycle in $\langle \zeta \rangle$, we have that h is $\mathrm{Sp}_4(q)$-conjugate to a matrix of the form (5.2).

Clearly,

$$\mathbf{C}_{\mathrm{Sp}_4(q)}(h_{\ell_1,\ell_2}) \geq \left\{ \begin{pmatrix} \zeta^i & 0 \\ 0 & \zeta^j \end{pmatrix} \mid i, j \in \{1, \ldots, q+1\} \right\} \quad (5.3)$$

and the group on the right hand side has order $(q+1)^2$. This proves our claims (†) and (††).

Let now, for simplicity, be $h_\ell := h_{1,\ell}$ and consider

$$\mathcal{E} := \left\{ h_\ell = \begin{pmatrix} \zeta & 0 \\ 0 & \zeta^\ell \end{pmatrix} \in \mathrm{Sp}_4(q) \mid \ell \in \{1, \ldots, q\}, \gcd(\ell, q+1) = 1 \right\}.$$

We claim that a matrix $h_\ell \in \mathcal{E}$ has centralizer of size $(q+1)^2$ if and only if $\ell \notin \{1, q\}$. In order to show that, we identify the natural module $\mathbb{F}_q^2$ for $\mathrm{Sp}_2(q)$ with $\mathbb{F}_{q^2}$ and take ζ to be the matrix representing the multiplication by the $(q-1)$th power of a generator of $\mathbb{F}_{q^2}^*$, again denoted by ζ. When $\ell \in \{1, q\}$, we can find $A, B \in \mathrm{Sp}_2(q)$ with $\zeta^\ell = A^{-1}\zeta A$ and $\zeta^\ell = B\zeta B^{-1}$. Indeed, when $\ell = 1$ this is obvious because we may take $A = B = I$, and when $\ell = q$ we may take the matrix B induced by the action on $\mathbb{F}_{q^2}$ of the Galois group of $\mathbb{F}_{q^2}/\mathbb{F}_q$ and $A = B^{-1}$. Then, the matrix

$$\begin{pmatrix} 0 & A \\ B & 0 \end{pmatrix}$$

lies in $\mathbf{C}_{\mathrm{Sp}_4(q)}(h_\ell)$ and hence $|\mathbf{C}_{\mathrm{Sp}_4(q)}(h_\ell)| > (q+1)^2$.

Conversely, let $\ell \in \{1, \ldots, q\} \setminus \{1, q\}$ with $\gcd(\ell, q+1) = 1$. To prove that the equality in (5.3) is attained it suffices to show that $\langle e_1, e_2 \rangle$ and $\langle e_3, e_4 \rangle$ are the only proper $\mathbb{F}_q\langle h_\ell \rangle$-submodules of $V = \langle e_1, e_2, e_3, e_4 \rangle$. Suppose that W is a proper $\mathbb{F}_q\langle h_\ell \rangle$-submodule different from $\langle e_1, e_2 \rangle$ and $\langle e_3, e_4 \rangle$. Observe that W is a 2-dimensional irreducible $\mathbb{F}_q\langle h_\ell \rangle$-submodule because h_ℓ has no eigenvalues in $\mathbb{F}_q$. Using the decomposition $V = \langle e_1, e_2 \rangle \oplus \langle e_3, e_4 \rangle$, for the computations that follow, we identify the relevant matrices of $\mathrm{Sp}_4(q)$ as 2×2 block matrices and the vectors of $\mathbb{F}_q^4$ as block vectors $(x, y) \in (\mathbb{F}_q^2)^2 \cong (\mathbb{F}_{q^2})^2$. Let $w \in W \setminus \{0\}$. By the irreducibility

5.1 Small Dimensional Symplectic Groups

of $\langle e_1, e_2 \rangle$ and of $\langle e_3, e_4 \rangle$, we have $W \cap \langle e_1, e_2 \rangle = 0$ and $W \cap \langle e_3, e_4 \rangle = 0$. Thus we can write

$$w = (x, y),$$

with $x \in \langle e_1, e_2 \rangle \setminus \{0\}$ and $y \in \langle e_3, e_4 \rangle \setminus \{0\}$. We identify then x and y with elements in $\mathbb{F}_{q^2}^*$.

Since W has dimension 2 and is irreducible, we have

$$W = \mathrm{Span}_{\mathbb{F}_q}\{w, h_\ell w\} = \mathrm{Span}_{\mathbb{F}_q}\left\{(x, y), (\zeta x, \zeta^\ell y)\right\}.$$

Thus $h_\ell^2 w \in \langle w, h_\ell w \rangle$ and hence, there exist $a, b \in \mathbb{F}_q$ with $h_\ell^2 w = aw + bh_\ell w$. We obtain the equalities

$$\zeta^2 x = ax + b\zeta x,$$
$$\zeta^{2\ell} y = ay + b\zeta^\ell y$$

in the field $\mathbb{F}_{q^2}$. By canceling x and y, we deduce $\zeta^2 - b\zeta - a = 0$ and $\zeta^{2\ell} - b\zeta^\ell - a = 0$. Therefore, ζ and ζ^ℓ are roots of the polynomial $X^2 - bX - a \in \mathbb{F}_q[x]$. Since the Galois group $\mathrm{Gal}(\mathbb{F}_{q^2}/\mathbb{F}_q)$ is cyclic of order 2 generated by $\lambda \mapsto \lambda^q$, we deduce that the roots of $X^2 - bX - a$ are ζ and ζ^q. Therefore $\zeta^\ell \in \{\zeta, \zeta^q\}$ and $\ell \in \{1, q\}$, which is a contradiction.

We are now ready to prove part (1). Indeed when $f = 1$, we have $q = 2$ and by (††) every $h \in \mathrm{Sp}_4(2)$ with $\mathbf{o}(h) = 3$ and centralizer of order 9 is conjugate to a matrix of type

$$h_\ell = \begin{pmatrix} \zeta & 0 \\ 0 & \zeta^\ell \end{pmatrix}$$

for a suitable Singer cycle $\zeta \in \mathrm{Sp}_2(2)$ and with $\ell \in \{1, 2\}$. But we have seen above that those matrices have centralizer of size greater than 9.

Let next $f > 1$. We then have

$$\mathcal{D} := \{h_\ell \in \mathcal{E} \mid \ell \in \{1, \ldots, q\} \setminus \{1, q\}\} \neq \emptyset$$

and, by the above considerations, $h_\ell \in \mathcal{D}$ implies $\mathbf{o}(h_\ell) = q+1$ and $|\mathbf{C}_{\mathrm{Sp}_4(q)}(h_\ell)| = (q+1)^2$.

We are now ready to prove part (2). We explore the normalizers of $\langle h_\ell \rangle$ for $h_\ell \in \mathcal{D}$. Inside $\mathrm{Sp}_2(q)$ the normalizer of the Singer cycle $\langle \zeta \rangle$ has order $2(q+1)$, where the "2" arises (as described above) from a Galois action. In particular, $\zeta^\eta \in \{\zeta, \zeta^q = \zeta^{-1}\}$, for every $\eta \in \mathbf{N}_{\mathrm{Sp}_2(q)}(\langle \zeta \rangle)$. From this, it follows that

$$|\mathbf{N}_{\mathrm{Sp}_2(q) \perp \mathrm{Sp}_2(q)}(\langle h_\ell \rangle) : \mathbf{C}_{\mathrm{Sp}_4(q)}(h_\ell)| = 2.$$

Since the elements in $\mathbf{N}_{\mathrm{Sp}_4(q)}(\langle h_\ell \rangle)$ permute the two subspaces left invariant by h_ℓ, we deduce

$$|\mathbf{N}_{\mathrm{Sp}_4(q)}(\langle h_\ell \rangle) : \mathbf{N}_{\mathrm{Sp}_2(q) \perp \mathrm{Sp}_2(q)}(h_\ell)| \leq 2$$

and hence (2) follows.

Before proving part (3) and (4), we first claim that

(†††) Let $\ell \in \{1, \ldots, q\} \setminus \{1, q\}$. Then $h_\ell \in \mathcal{D}$ satisfies $|\mathbf{N}_{\mathrm{Sp}_4(q)}(\langle h_\ell \rangle) : \mathbf{C}_{\mathrm{Sp}_4(q)}(h_\ell)| = 4$ if and only if $\ell^2 \equiv \pm 1 \bmod (q+1)$.

From above, $|\mathbf{N}_{\mathrm{Sp}_4(q)}(\langle h_\ell \rangle) : \mathbf{C}_{\mathrm{Sp}_4(q)}(h_\ell)| = 4$ if and only if there exists a matrix

$$g := \begin{pmatrix} 0 & X \\ Y & 0 \end{pmatrix} \in \mathbf{N}_{\mathrm{Sp}_4(q)}(\langle h_\ell \rangle),$$

for suitable 2×2 matrices X and Y in $\mathrm{Sp}_2(q)$. A computation yields that $Y^{-1} \zeta^\ell Y, X^{-1} \zeta X \in \langle \zeta \rangle = \langle \zeta^\ell \rangle$ and hence $Y^{-1} \zeta Y, X^{-1} \zeta X \in \{\zeta, \zeta^{-1}\}$. This shows that $g^{-1} h_\ell g$ is one of the following four matrices

$$\begin{pmatrix} \zeta^\ell & 0 \\ 0 & \zeta \end{pmatrix}, \begin{pmatrix} \zeta^\ell & 0 \\ 0 & \zeta^{-1} \end{pmatrix}, \begin{pmatrix} \zeta^{-\ell} & 0 \\ 0 & \zeta \end{pmatrix}, \begin{pmatrix} \zeta^{-\ell} & 0 \\ 0 & \zeta^{-1} \end{pmatrix}.$$

An easy computation shows that at least one of these matrices is in $\langle h_\ell \rangle$ if and only if $\ell^2 \equiv \pm 1 \bmod(q+1)$.

To prove part (3) observe that when $q = 4$ every element ℓ in $\{1, \ldots, q\} \setminus \{1, q\}$ with $\gcd(\ell, q+1) = 1$ satisfies $\ell^2 \equiv \pm 1 \bmod(q+1)$; however, when $q > 4$, we may always choose ℓ in $\{1, \ldots, q\} \setminus \{1, q\}$ with $\gcd(\ell, q+1) = 1$ and $\ell^2 \not\equiv \pm 1 \bmod(q+1)$.

Finally, we show part (4). Let f be even and let $h \in \mathcal{D}$ be $\mathrm{Sp}_4(q)$-conjugate to an element x of the maximal subgroup $\mathrm{Sp}_4(\sqrt{q}) \in \mathcal{C}_5$. Then $\mathbf{o}(x) = q+1 = (\sqrt{q})^2 + 1$ and, by Proposition 2.5, x is a Singer cycle of $\mathrm{Sp}_4(\sqrt{q})$. Thus $|\mathbf{N}_{\mathrm{Sp}_4(\sqrt{q})}(\langle x \rangle) : \langle x \rangle| = 4$, where the "4" arises from the Galois action of $\mathrm{Gal}(\mathbb{F}_{(\sqrt{q})^4}/\mathbb{F}_{\sqrt{q}})$. Then

$$4 \mid |\mathbf{N}_{\mathrm{Sp}_4(q)}(\langle h \rangle) : \langle h \rangle| = |\mathbf{N}_{\mathrm{Sp}_4(q)}(\langle h \rangle) : \mathbf{C}_{\mathrm{Sp}_4(q)}(h)|(q+1).$$

Since q is even, this implies $4 \mid |\mathbf{N}_{\mathrm{Sp}_4(q)}(\langle h \rangle) : \mathbf{C}_{\mathrm{Sp}_4(q)}(h)|$ and thus, by (2), we get $|\mathbf{N}_{\mathrm{Sp}_4(q)}(\langle h \rangle) : \mathbf{C}_{\mathrm{Sp}_4(q)}(h)| = 4$ which proves part (4). □

Lemma 5.4 *Let f be a positive integer with $f \geq 2$ and let $q := 2^f$. The weak normal covering number of $\mathrm{Sp}_4(q) = \mathrm{PSp}_4(q)$ is 2. Moreover, if H and K are maximal components of a weak normal 2-covering of $\mathrm{Sp}_4(q)$, then up to $\mathrm{Aut}(\mathrm{Sp}_4(q))$-conjugacy one of the following holds*

5.1 Small Dimensional Symplectic Groups

(1) $H \cong SO_4^-(q)$ and $K \cong SO_4^+(q)$ are in class $\mathcal{C}_8$. In this case the weak normal covering gives rise, up to $Sp_4(q)$-conjugacy, to the two normal coverings having components

  (a) $H \cong SO_4^-(q)$ and $K \cong SO_4^+(q)$;
  (b) $H \cong Sp_2(q^2) : 2 \in \mathcal{C}_3$ and $K \cong Sp_2(q) \text{wr} \, 2 \in \mathcal{C}_2$.

(2) $q = 4$, $H \cong Sp_2(16) : 2 \in \mathcal{C}_3$ and $K \cong Sp_4(2) \in \mathcal{C}_5$. In this case the weak normal covering does not give rise to normal coverings.

In particular, $\gamma(Sp_4(q)) = 2$.

Proof When $q = 4$, the proof follows with a computer computation. Therefore, for the rest of the proof, we assume that $q > 4$.

Let H be a maximal component of a weak normal covering μ of $Sp_4(q)$ of minimum size, with H containing a Singer cycle. From Lemma 5.1, up to $Sp_4(q)$-conjugacy, we have

$$H \cong SO_4^-(q) \in \mathcal{C}_8 \text{ or } H \cong Sp_2(q^2) : 2 \in \mathcal{C}_3.$$

Observe that a graph automorphism γ of $Sp_4(q)$ fuses $SO_4^-(q)$ and $Sp_2(q^2) : 2$, see [9, Table 8.14]. Therefore, replacing H with a suitable $Aut(Sp_4(q))$-conjugate, we may suppose that $H \cong SO_4^-(q)$. From Lemma 5.3, the group $Sp_4(q)$ contains an element y having order $q + 1$ and with $|C_{Sp_4(q)}(y)| = (q + 1)^2$. Now assume, by contradiction, that $H \cong SO_4^-(q) \cong SL_2(q^2) : 2$ contains an element y with $o(y) = q + 1$ and $|C_{Sp_4(q)}(y)| = (q + 1)^2$. Since $q^2 - 1 \mid |C_{SL_2(q^2)}(y)|$, we then have $q^2 - 1 \mid (q + 1)^2$ and thus $q - 1 \mid q + 1$, against $\gcd(q - 1, q + 1) = 1$. Hence we deduce that μ contains a second maximal component K, with $y \in K$. A case-by-case analysis on the maximal subgroups of $Sp_4(q)$ in [9, Table 8.14] reveals that, up to $Sp_4(q)$-conjugacy, there are three possibilities for K. Namely

  (i) $K \cong Sp_2(q) \text{wr} \, 2 \in \mathcal{C}_2$,
  (ii) $K \cong SO_4^+(q) \in \mathcal{C}_8$,
  (iii) f is even and $K \cong Sp_4(\sqrt{q}) \in \mathcal{C}_5$.

Since the graph automorphism γ of $Sp_4(q)$ fuses $SO_4^+(q)$ and $Sp_2(q) \text{wr} \, 2$, up to $Aut(Sp_4(q))$-conjugacy there are only two possibilities for K.

Consider first the case f even and $K \cong Sp_4(\sqrt{q}) \in \mathcal{C}_5$. From Lemma 5.3, we see that, when $q > 4$, not every element of order $q + 1$ and centralizer of cardinality $(q + 1)^2$ is $Aut(Sp_4(q))$-conjugate to an element of K. Thus H, K are not components of a weak normal 2-covering. Thus we exclude from any further consideration the choice $K \cong Sp_4(\sqrt{q}) \in \mathcal{C}_5$.

The choice $K \cong SO_4^+(q)$ leads, by [33], to a weak normal covering which is also a normal covering. Moreover, since γ simultaneously fuses $SO_4^-(q)$ and $Sp_2(q^2) : 2$ as well as $SO_4^+(q)$ and $Sp_2(q) \text{wr} \, 2$, the weak normal covering in (1) gives rise to at least the two distinct normal coverings (a) and (b).

At this point, it suffices to show that the subgroups $SO_4^-(q)$, $Sp_2(q)$wr2 are not the components of a normal 2-covering. Indeed, once this is established, the same conclusion follows for the subgroups $SO_4^+(q)$, $Sp_2(q^2) : 2$. Now let s be a Singer cycle of $GL_2(q)$ and consider the matrix

$$x := \begin{pmatrix} s & 0 \\ 0 & (s^{-1})^T \end{pmatrix}.$$

The matrix x lies in $Sp_4(q)$ with respect to the non-degenerate symplectic form defined by the matrix

$$J = \begin{pmatrix} 0 & 0 & 1 & 0 \\ 0 & 0 & 0 & 1 \\ 1 & 0 & 0 & 0 \\ 0 & 1 & 0 & 0 \end{pmatrix},$$

has order $q^2 - 1$ and action $2 \oplus 2$. In particular, x has no eigenvalues in the ground field $\mathbb{F}_q$.

Assume, by contradiction, that x is $Sp_4(q)$-conjugate to $y \in Sp_2(q)$wr$2 \in \mathcal{C}_2$. Since q is even, then y lies in the base group $Sp_2(q) \perp Sp_2(q)$ so that y has the form $s_1 \perp s_2$, where $s_1, s_2 \in Sp_2(q)$ with $\langle s_i \rangle$ irreducible cyclic subgroups of $Sp_2(q)$. Thus $o(s_i) \mid q + 1$ for $i \in \{1, 2\}$ and hence also $q^2 - 1 = o(y) \mid q + 1$, against $q > 4$.

Assume next, by contradiction, that x is $Sp_4(q)$-conjugate to $y \in SO_4^-(q) \in \mathcal{C}_8$. Observe that the quadratic form $Q : \mathbb{F}_q^4 \to \mathbb{F}_q$ defined by $Q(a, b, c, d) = ac + bd$ is non-degenerate, polarizes to J and has Witt defect zero. Since x preserves Q, we deduce that $x \in SO_4^+(q)$, where $SO_4^+(q)$ is defined by Q. Let $\pi : Sp_4(q) \to \mathbb{C}$ be the permutation character for the action of $Sp_4(q)$ on $V = \mathbb{F}_q^4$. Similarly, let $\pi^+ : Sp_4(q) \to \mathbb{C}$ and $\pi^- : Sp_4(q) \to \mathbb{C}$ be the permutation characters for the action of $Sp_4(q)$ on the right cosets of $SO_4^+(q)$ and of $SO_4^-(q)$ respectively. It is shown in [50, Theorem 1] that $\pi = \pi^+ + \pi^-$. Now, $\pi(y) = \pi(x) = |C_V(x)| = 1$, because x fixes only the zero vector of V. Since $x \in SO_4^+(q)$, we have $\pi^+(y) = \pi^+(x) \geq 1$; therefore, $\pi^+(y) = 1$ and $\pi^-(y) = 0$. As $\pi^-(y) = 0$, y has no $Sp_4(q)$-conjugate in $SO_4^-(q)$. □

We now examine the group $PSp_6(q)$.

Lemma 5.5 *The weak normal covering number of $PSp_6(q)$ is at least 2. Moreover, if H and K are maximal components of a weak normal 2-covering of $PSp_6(q)$, then up to $\mathrm{Aut}(PSp_6(q))$-conjugacy one of the following holds*

(1) q is even, $H \cong SO_6^-(q)$ and $K \cong SO_6^+(q)$ are in class $\mathcal{C}_8$,
(2) $q = 3^f$ for some $f \geq 1$, $\tilde{H} \cong Sp_2(q) \perp Sp_4(q) \in \mathcal{C}_1$ and $\tilde{K} \cong Sp_2(q^3) : 3 \in \mathcal{C}_3$.

The coverings in (1) and (2) give both rise to a single normal covering.

5.1 Small Dimensional Symplectic Groups

Proof When $q \in \{2, 3, 4\}$, the result follows with a computation with the computer algebra system magma [6]. Therefore, for the rest of the argument, we suppose $q \geq 5$.

Set $\kappa := \gamma_w(\mathrm{PSp}_6(q))$ and let H be a maximal component of a weak normal κ-covering μ of $\mathrm{PSp}_6(q)$, with H containing a Singer cycle x. From Lemma 5.1, up to $\mathrm{Sp}_6(q)$-conjugacy, one of the following holds

(i) $\tilde{H} \cong \mathrm{Sp}_2(q^3) : 3 \in \mathcal{C}_3$,
(ii) $\tilde{H} \cong \mathrm{GU}_3(q).2 \in \mathcal{C}_3$, with q odd,
(iii) $\tilde{H} \cong \mathrm{SO}_6^-(q) \in \mathcal{C}_8$, with q even.

From [9, Tables 8.28, 8.29], we see that there exist two $\mathrm{Sp}_6(q)$-conjugacy classes of maximal subgroups containing an element having order

$$(q+1)(q^2+1)/\gcd(q+1, q^2+1) = (q+1)(q^2+1)/\gcd(2, q-1)$$

and these are as follows

(a) $\mathrm{Sp}_2(q) \perp \mathrm{Sp}_4(q) \in \mathcal{C}_1$,
(b) $\mathrm{SO}_6^+(q) \in \mathcal{C}_8$, with q even.

This already shows that $\kappa \geq 2$, because none of the possibilities for $\tilde{H}$ is in this list. Assume now that there exists a weak normal 2-covering of $\mathrm{PSp}_6(q)$ with maximal components. By the above arguments, the corresponding weak normal 2-covering of $\mathrm{Sp}_6(q)$ is given by $\tilde{\mu} = \{\tilde{H}, \tilde{K}\}$, where $\tilde{H}$ must be in one of the three possibilities (i)–(iii) and $\tilde{K}$ must be in one of the two possibilities (a)–(b).

Let z be an element of $\mathrm{Sp}_6(q)$ having order $q^3 - 1$. From [9, Tables 8.28, 8.29] and from the fact that $q \geq 5$, we see that there exist four $\mathrm{Sp}_6(q)$-conjugacy classes of maximal subgroups containing an element z having order $q^3 - 1$ and these are as follows

(α) $E_q^6 : \mathrm{GL}_3(q) \in \mathcal{C}_1$,
(β) $\mathrm{Sp}_2(q^3) : 3 \in \mathcal{C}_3$,
(γ) $\mathrm{GL}_3(q).2 \in \mathcal{C}_2$, with q odd,
(δ) $\mathrm{SO}_6^+(q) \in \mathcal{C}_8$, with q even.

Assume first that $\tilde{K}$ is of type

$$\mathrm{Sp}_2(q) \perp \mathrm{Sp}_4(q).$$

Now, $\tilde{K}$ contains no elements of order $q^3 + 1$ or $q^3 - 1$ and hence x and z are $\mathrm{Aut}(\mathrm{Sp}_6(q))$-conjugate to elements in $\tilde{H}$. By consulting the possibilities in the lists (i)–(iii) and (α)–(γ), we deduce that $\mathrm{Sp}_2(q^3) : 3$ is the only possible group in both lists and hence $\tilde{H}$ is of type

$$\mathrm{Sp}_2(q^3) : 3.$$

Write $q = p^f$, where p is a prime number and $f \geq 1$. Suppose that $p \neq 3$. Observe that the non-identity unipotent elements of $\tilde{K}$ centralize a subspace of dimension at least 2 of V, whereas the non-identity unipotent elements of $\tilde{H}$ are contained in $\mathrm{Sp}_2(q^3)$ because $p \neq 3$ and hence centralize a subspace of dimension at least 3 of V. As $\mathrm{Sp}_6(q)$ has unipotent elements u with $\dim_{\mathbb{F}_q} V_1(u) = 1$, as can be easily verified considering the possible Jordan forms of a 6×6 symplectic matrix over $\mathbb{F}_q$, u is not $\mathrm{Aut}(\mathrm{Sp}_6(q))$-conjugate to an element of $\tilde{H}$ or of $\tilde{K}$. Hence $\tilde{H}, \tilde{K}$ are not components of a normal 2-covering. This contradiction says that we have

$$p = 3,\ \tilde{H} \cong \mathrm{Sp}_2(q^3) : 3 \text{ and } \tilde{K} \cong \mathrm{Sp}_2(q) \perp \mathrm{Sp}_4(q),\ \text{or } q \text{ is even and } K \cong \mathrm{SO}_6^+(q).$$

When q is a power of 3, we obtain the weak normal covering in (2). This is indeed a genuine normal 2-covering of $\mathrm{Sp}_6(q)$, see [16].

Assume now q even and $K \cong \mathrm{SO}_6^+(q)$. Now, K contains no elements of order $q^3 + 1$ and hence x is $\mathrm{Aut}(\mathrm{Sp}_6(q))$-conjugate to an element in H. By consulting the possibilities above, we deduce that H is of type $\mathrm{Sp}_2(q^3) : 3$ or $\mathrm{SO}_6^-(q)$. In the second case, we get the normal covering in (1), by [33]. Assume now that H is of type $\mathrm{Sp}_2(q^3) : 3$. Let $w \in \mathrm{Sp}_6(q)$ having order $(q - 1)(q^2 + 1)$ and action type $1 \oplus 1 \oplus 4$. Clearly, w cannot be $\mathrm{Aut}(\mathrm{Sp}_6(q))$-conjugate to an element of $K \cong \mathrm{Sp}_2(q^3) : 3$. Suppose, by contradiction, that w is $\mathrm{Aut}((\mathrm{Sp}_6(q))$-conjugate to an element $w' \in \mathrm{SO}_6^+(q)$ and let Q be the quadratic form preserved by $\mathrm{SO}_6^+(q)$. Now, w' preserves an orthogonal direct sum decomposition $V_1 \oplus V_2 \oplus V_3$ of $\mathbb{F}_q^6$ with $\dim_{\mathbb{F}_q}(V_1) = \dim_{\mathbb{F}_q}(V_2) = 1$ and $\dim_{\mathbb{F}_q}(V_3) = 4$. Since Q is non-degenerate, V_3 cannot be totally isotropic with respect to Q and hence the restriction Q' of Q to V_3 induces a non-degenerate quadratic form. The action of w' on V_3 induces a Singer cycle and hence the orthogonal group defined by Q' contains a Singer cycle. By Sect. 2.5, Q' has Witt defect 1. Since Q has Witt defect 0, we deduce that the restriction Q'' of Q to $V_1 \oplus V_2$ has Witt defect 1. In particular, the isometry group of Q'' is $\mathrm{SO}_2^-(q)$. Therefore, the action of w' on $V_1 \oplus V_2$ induces a matrix in $\mathrm{SO}_2^-(q) \cong D_{2(q+1)}$ of order $q - 1$ and of type $1 \oplus 1$, which is a contradiction. Thus the pair $\mathrm{SO}_6^+(q), \mathrm{Sp}_2(q^3) : 3$ does not give rise to a weak normal 2-covering. $\square$

Finally, we deal with 8-dimensional symplectic groups.

Lemma 5.6 *The weak normal covering number of* $\mathrm{PSp}_8(q)$ *is at least 2. Moreover, if H and K are maximal components of a weak normal 2-covering of* $\mathrm{PSp}_8(q)$, *then q is even and up to* $\mathrm{Aut}(\mathrm{Sp}_8(q))$-*conjugacy* $H \cong \mathrm{SO}_8^+(q)$ *and* $K \cong \mathrm{SO}_8^-(q)$ *are in class* $\mathcal{C}_8$. *This covering gives rise to a single normal covering.*

Proof When $q \in \{2, 3\}$, the result follows with a computation with the computer algebra system magma [6]. Therefore, for the rest of the argument, we suppose $q \geq 4$.

Set $\kappa := \gamma_w(\mathrm{PSp}_8(q))$ and let H be a maximal component of a weak normal κ-covering μ, with H containing a Singer cycle x. From Lemma 5.1, up to $\mathrm{Sp}_8(q)$-conjugacy, $\tilde{H}$ is one of the following

(i) $\mathrm{Sp}_4(q^2) : 2 \in \mathcal{C}_3$,
(ii) $\mathrm{SO}_8^-(q) \in \mathcal{C}_8$, with q even.

5.1 Small Dimensional Symplectic Groups

Observe that, from [9, Tables 8.48, 8.49], we infer that in both possibilities[3]

$$\text{Aut}(\tilde{G}) = \mathbf{N}_{\text{Aut}(\tilde{G})}(\tilde{H})\tilde{G}. \tag{5.4}$$

This already shows that $\kappa \geq 2$.

From [9, Tables 8.48, 8.49] and from $q \geq 4$, we see that the $\text{Sp}_8(q)$-conjugacy classes of maximal subgroups containing an element having order

$$\frac{(q+1)(q^3-1)}{\gcd(q+1, q^3-1)} = \frac{(q+1)(q^3-1)}{\gcd(2, q-1)}$$

are as follows

(a) $\text{Sp}_2(q) \perp \text{Sp}_6(q) \in \mathcal{C}_1$,
(b) $E_q^{6+6} : (\text{GL}_3(q) \times \text{Sp}_2(q)) \in \mathcal{C}_1$,
(c) $\text{SO}_8^-(q)$, with q even, in class $\mathcal{C}_8$.

Let now μ be a weak normal 2-covering of $\text{PSp}_8(q)$ with maximal components. By the above arguments, the corresponding weak normal 2-covering of $\text{Sp}_8(q)$ is given by $\tilde{\mu} = \{\tilde{H}, \tilde{K}\}$, where $\tilde{H}$ must be in one of the two possibilities (i)–(ii) and $\tilde{K}$ must be in one of the three possibilities (a)–(c).

From [9, Tables 8.48, 8.49] and from $q \geq 4$, we see that the $\text{Sp}_8(q)$-conjugacy classes of maximal subgroups containing an element having order $q^4 - 1$ and having type $4 \oplus 4$ are as follows

(1) $E_q^{10} : \text{GL}_4(q) \in \mathcal{C}_1$,
(2) $\text{Sp}_4(q^2) : 2 \in \mathcal{C}_3$,
(3) $\text{GL}_4(q).2$, with q odd, in class $\mathcal{C}_2$,
(4) $\text{GU}_4(q).2$, with q odd, in class $\mathcal{C}_3$,
(5) $\text{SO}_8^+(q)$, with q even, in class $\mathcal{C}_8$.

This is not hard to check, but rather tedious. To exclude the subgroups in the Aschbacher class $\mathcal{S}$ it is more convenient and time efficient to observe that they do not contain elements having order $q^4 - 1$. Moreover, when q is even, to exclude the subgroups $E_q^7 : ((q-1) \times \text{Sp}_6(q))$, $E_q^{3+8} : (\text{GL}_2(q) \times \text{Sp}_4(q))$ and $\text{Sp}_2(q) \perp \text{Sp}_6(q)$ in the Aschbacher class $\mathcal{C}_1$, it suffices to observe that none of the elements having order $q^4 - 1$ in these groups have type $4 \oplus 4$.

Assume first that

$$\tilde{K} \cong E_q^{6+6} : (\text{GL}_3(q) \times \text{Sp}_2(q)).$$

As $\tilde{K}$ contains no elements of order $q^4 + 1$ or $q^4 - 1$, $\tilde{H}$ contains elements having order $q^4 + 1$ and $q^4 - 1$. By consulting the lists of options (i)–(ii) and (1)–(5), we

---

[3] Note that in [9], the normalizer is denoted with "Stab".

deduce that $\tilde{H}$ is of type $\mathrm{Sp}_4(q^2) : 2$. Now, consider elements $t \in \mathrm{Sp}_8(q)$ with $\mathrm{o}(t) = (q-1)(q^3+1)/\gcd(q-1, q^3+1) = (q-1)(q^3+1)/\gcd(2, q-1)$. Since $E_q^{6+6} : (\mathrm{GL}_3(q) \times \mathrm{Sp}_2(q))$ and $\mathrm{Sp}_4(q^2) : 2$ do not contain elements having this order, we deduce that t is contained in no $\mathrm{Aut}(\mathrm{Sp}_8(q))$-conjugate of $\tilde{H}$ or $\tilde{K}$. Therefore, a weak normal 2-covering cannot make use of $E_q^{6+6} : (\mathrm{GL}_3(q) \times \mathrm{Sp}_2(q))$.

Assume now that

$$\tilde{K} = \mathrm{Sp}_2(q) \perp \mathrm{Sp}_6(q).$$

Therefore, $\tilde{H}$ is of type $\mathrm{Sp}_4(q^2) : 2$, or $\mathrm{SO}_8^-(q)$ with q even. By considering the unipotent elements of $\mathrm{Sp}_8(q)$, we see that when q is odd $\tilde{H}$ cannot be $\mathrm{Sp}_4(q^2) : 2$. Thus q is even. Now, consider the non-degenerate symplectic form on $\mathbb{F}_q^4$ having matrix

$$J := \begin{pmatrix} 0 & 0 & 0 & 1 \\ 0 & 0 & 1 & 0 \\ 0 & 1 & 0 & 0 \\ 1 & 0 & 0 & 0 \end{pmatrix}$$

and let $\mathrm{Sp}_4(q)$ be the symplectic group with respect to this form. Now, using the matrix J, it can be easily verified that

$$u := \begin{pmatrix} 1 & 1 & 1 & 0 \\ 0 & 1 & 1 & 0 \\ 0 & 0 & 1 & 1 \\ 0 & 0 & 0 & 1 \end{pmatrix} \in \mathrm{Sp}_4(q)$$

and that u has order 4. Moreover, u has a unique u-invariant 2-dimensional subspace (namely, the subspace spanned by the last two basis vectors) and this subspace is totally isotropic with respect to J. Now, let s be a Singer cycle of $\mathrm{Sp}_4(q)$ and set

$$g := u \oplus s \in \mathrm{Sp}_8(q).$$

Clearly, $\mathrm{o}(g) = 4(q^2 + 1)$. Now, the element g has no $\mathrm{Aut}(\mathrm{Sp}_8(q))$-conjugate in $\tilde{K} = \mathrm{Sp}_2(q) \perp \mathrm{Sp}_6(q)$ because g fixes no non-degenerate 2-subspace of V. Suppose that g has an $\mathrm{Aut}(\mathrm{Sp}_8(q))$-conjugate in $\mathrm{Sp}_4(q^2) : 2$ and recall that $\mathrm{Sp}_4(q^2) : 2$ is a field extension subgroup. Without loss of generality, we may suppose that $g \in \mathrm{Sp}_4(q^2) : 2$. Now, when we see $g = u \oplus s$ as an element of $\mathrm{Sp}_4(q^2) : 2$, s becomes a semisimple element fixing a 2-dimensional non-degenerate subspace over $\mathbb{F}_{q^2}$. From this, we get

$$\mathbf{C}_{\mathrm{Sp}_4(q^2)}(s) \cong \mathrm{Sp}_2(q^2) \perp \langle s \rangle.$$

5.1 Small Dimensional Symplectic Groups

Observe that the centralizer of s in $\mathrm{Sp}_4(q^2) : 2$ is contained in $\mathrm{Sp}_4(q^2)$, because the ":2" on top acts via Galois conjugation and hence it cannot centralize the matrix s being a Singer cycle of $\mathrm{Sp}_2(q^2)$. Thus

$$u \in \mathbf{C}_{\mathrm{Sp}_4(q^2):2}(s) = \mathbf{C}_{\mathrm{Sp}_4(q^2)}(s) \subseteq \mathrm{Sp}_2(q^2) \perp \langle s \rangle$$

and hence $u \in \mathrm{Sp}_2(q^2) \cong \mathrm{SL}_2(q^2)$. However, non-identity unipotent elements of $\mathrm{SL}_2(q^2)$ have order 2 and not 4. This contradiction shows that the element g has no $\mathrm{Aut}(\mathrm{Sp}_8(q))$-conjugate in $\mathrm{Sp}_4(q^2) : 2$. It remains to consider the case that $\tilde{H} \cong \mathrm{SO}_8^-(q)$. For dealing with this case, we turn to an element y having order $q^4 - 1$ and type $4 \oplus 4$. Since neither $\tilde{H}$ nor $\tilde{K}$ is one of the groups in (1)–(5), we deduce that y has no $\mathrm{Aut}(\mathrm{Sp}_8(q))$-conjugate in $\tilde{H}$ or in $\tilde{K}$. Hence this choice of $\tilde{H}$ and $\tilde{K}$ does not give rise to a weak normal 2-covering. This whole paragraph has shown that a weak normal 2-covering cannot make use of $\mathrm{Sp}_2(q) \perp \mathrm{Sp}_6(q)$.

The only remaining possibility is that

$$\tilde{K} \cong \mathrm{SO}_8^-(q) \text{ and } q \text{ even}.$$

Since $\mathrm{Sp}_8(q)$ and $\mathrm{SO}_8^-(q)$ are isospectral[4] we need to argue in more detail. Observe that the group $\tilde{H}$ appears in (1)–(5) and, since q is even, we may exclude (3) and (4). Thus $\tilde{H}$ is one of the following groups:

$$\mathrm{Sp}_4(q^2) : 2, \ E_q^{10} : \mathrm{GL}_4(q) \text{ and } \mathrm{SO}_8^+(q). \tag{5.5}$$

Finally we turn to the elements having order $q^3 + 1$ in $\mathrm{Sp}_8(q)$ and type $2 \oplus 6$. The elements having type $2 \oplus 6$ are not conjugate to an element of $\mathrm{SO}_8^-(q)$ because they do not preserve a non-degenerate quadratic form of "minus type".[5] Therefore the elements of order $q^3 + 1$ and of type $2 \oplus 6$ are all $\mathrm{Aut}(\mathrm{Sp}_8(q))$-conjugate to elements of $\tilde{H}$. Using elementary order considerations, from (5.5), we see that the

---

[4] Let X be a finite group. The spectrum of X is the set $\{\mathbf{o}(x) \mid x \in X\}$ consisting of the orders of the elements of X. Two finite groups X and Y are said to be isospectral if they have the same spectrum. The fact that $\mathrm{Sp}_8(q)$ and $\mathrm{SO}_8^-(q)$ are isospectral follows from [24, Corollary 3 and Theorem 4, p. 195].

[5] Let x be an element of $\mathrm{Sp}_8(q)$ having type $2 \oplus 6$ and order $q^3 + 1$. Then the vector space $V = \mathbb{F}_q^8$ decomposes as the sum of two irreducible $\mathbb{F}_q\langle x \rangle$-modules, say $\mathbb{F}_q^8 = V_1 \oplus V_2$ with $\dim_{\mathbb{F}_q}(V_1) = 2$ and $\dim_{\mathbb{F}_q}(V_2) = 6$. Suppose $x \in \mathrm{SO}_8^-(q)$. Let Q be the quadratic form defining $\mathrm{SO}_8^-(q)$. Since the action of x on V_1 and on V_2 restricts to a Singer cycle, from Sect. 2.5, we deduce that Q restricts to V_1 and to V_2 to a quadratic form having Witt defect 1. As V is the orthogonal sum of V_1 and V_2, we deduce that Q has Witt defect 0, which is a contradiction.

Observe that we have used a similar argument in the last paragraph of the proof of Lemma 5.5. We make this argument more general when we deal with orthogonal groups.

only possibility for $\tilde{H}$ is $SO_8^+(q)$. Now, by Dye [33], for q even, $\tilde{K} \cong SO_8^-(q)$ and $\tilde{H} \cong SO_8^+(q)$ do give rise to a normal covering of $Sp_8(q)$. □

5.2 Large Dimensional Symplectic Groups

In this section we deal with large dimensional symplectic groups $Sp_n(q)$ with $n \geq 10$, n even. We consider the Bertrand number t for the integer $n/2 \geq 5$ (see Sect. 2.10). Thus for $n \geq 16$ we have $\frac{n}{4} < t \leq \frac{n}{2} - 3$ and t is prime; for $n = 14$, we have $t = 5$; for $n = 12$, we have $t = 4$; for $n = 10$ we have $t = 3$. Note that, for every $n \geq 10$, $\frac{n}{4} < t \leq \frac{n}{2} - 2$ holds. Moreover, t is a prime, except when $n = 12$. Recall also that, for every $n \geq 10$, we have $t \nmid n/2$. Moreover, since $\gcd(n/2, t) = 1$ when $n \neq 12$ and $\gcd(n/2, t) = 2$ when $n = 12$, from Lemma 2.2 (2) we get

$$\gcd(q^t + 1, q^{\frac{n}{2}-t} + 1) = \begin{cases} \gcd(2, q-1) & \text{if } n = 12 \text{ or } n/2 \text{ odd}, \\ q + 1 & \text{if } n \neq 12 \text{ and } n/2 \text{ even}. \end{cases} \quad (5.6)$$

In order to find the components of a weak normal covering of $Sp_n(q)$, we consider a Bertrand element $z \in Sp_n(q)$ such that

- $o(z) = \frac{(q^t+1)(q^{\frac{n}{2}-t}+1)}{\gcd(q^t+1, q^{\frac{n}{2}-t}+1)}$ (see Table 2.2),
- the action of z on V is of type $2t \oplus (n-2t)$ and we write $V = U \perp W$, where $\dim_{\mathbb{F}_q} U = 2t$, $\dim_{\mathbb{F}_q} W = n - 2t$ and U, W are irreducible $\mathbb{F}_q\langle z \rangle$-modules,
- z induces a matrix of order $q^t + 1$ on U and of order $q^{\frac{n}{2}-t} + 1$ on W.

Note that the existence of Bertrand elements is an immediate consequence of the embedding of $Sp_{2t}(q) \perp Sp_{n-2t}(q)$ in $Sp_n(q)$.

Lemma 5.7 *If M is a maximal subgroup of $Sp_n(q)$ with $n \geq 10$ containing a Bertrand element, then one of the following holds*

(1) $M \cong Sp_{2t}(q) \perp Sp_{n-2t}(q) \in \mathcal{C}_1$,
(2) $n/2$ is even, $n \neq 12$, q is odd and $M \cong GU_{n/2}(q).2 \in \mathcal{C}_3$,
(3) q is even and $M \cong SO_n^+(q) \in \mathcal{C}_8$,
(4) $n = 12$ and $M \cong Sp_6(q^2).2 \in \mathcal{C}_3$.

Proof Observe that $P_{2t}(q) \neq \emptyset$ as long as $(t, q) \neq (3, 2)$, that is, $(n, q) \neq (10, 2)$. When $(n, q) = (10, 2)$, we have verified the veracity of this result with the help of a computer; the maximal subgroup M is isomorphic to either $Sp_6(2) \perp Sp_4(2)$ as in (1), or $SO_{10}^+(2)$ as in (3). Therefore, for the rest of the proof, we suppose that $(n, q) \neq (10, 2)$. From (5.6), we immediately see that

$$z \text{ is a strong } ppd(n, q; 2t)\text{-element.}$$

5.2 Large Dimensional Symplectic Groups

Since $2t \leq n - 4$, by Theorem 2.6, M belongs to one of the Aschbacher classes $\mathcal{C}_i$, with $i \in \{1, 2, 3, 5, 8\}$ or to $\mathcal{S}$ as described in Example 2.6 a) of [53]. Note that the class $\mathcal{C}_6$ is excluded using $2t \leq n - 4$, see the last assertion in Theorem 2.6. We now divide the proof depending on the Aschbacher class of M.

The Subgroup M Lies in Class $\mathcal{C}_1$ Here M is the stabilizer of a proper totally isotropic or non-degenerate subspace of V. As usual we use the notation in [66]. By definition of z, the only z-invariant subspaces of V are U and W. As U and W are non-degenerate and $\dim_{\mathbb{F}_q} U = 2t$, we deduce $M \cong \mathrm{Sp}_{2t}(q) \perp \mathrm{Sp}_{n-2t}(q)$.

The Subgroup M Lies in Class $\mathcal{C}_2$ From [66, Section 4.2], M is of type $\mathrm{Sp}_m(q) \mathrm{wr} S_\ell$ for some even divisor m of n with $n = m\ell$ and $\ell \geq 2$, or of type $\mathrm{GL}_{n/2}(q).2$ with q odd. The detailed structure of M is described in [66, Propositions 4.2.5 and 4.2.10].

Suppose M is of type $\mathrm{Sp}_m(q) \mathrm{wr} S_\ell$. Let $r \in P_{2t}(q)$. Since r divides the order of M and $\gcd(r, q) = 1$, we have that either r divides $q^{2i} - 1$, for some $i \in \{1, \ldots, m/2\}$, or $r \leq \ell$. In the first case, by Lemma 2.3 (3), we deduce $2t \mid 2i$ and thus $t \leq i \leq m/2 = n/(2\ell) \leq n/4$, contradicting the fact that $t > n/4$. In the second case, as r is a primitive prime divisor of $q^{2t} - 1$, from (2.4), we have $r > 2t$. Hence $\ell \geq r > 2t > n/2$, contradicting the fact that $\ell \leq n/2$.

Suppose M is of type $\mathrm{GL}_{n/2}(q).2$. Now, it is easily seen, arguing as in the previous paragraph, that no element in $P_{2t}(q)$ divides the order of M and hence this case does not arise.

The Subgroup M Lies in Class $\mathcal{C}_3$ From [66, Section 4.3], M is of type $\mathrm{Sp}_{n/r}(q^r).r$ for some prime divisor r of $n/2$ or of type $\mathrm{GU}_{n/2}(q).2$ with q odd. The detailed structure of M is described in [66, Propositions 4.3.7 and 4.3.10].

Suppose first that M is of type $\mathrm{Sp}_{n/r}(q^r).r$. If $n = 12$, then $t = 4$, $r \in \{2, 3\}$ and

$$\mathbf{o}(z) = \frac{(q^4 + 1)(q^2 + 1)}{\gcd(2, q - 1)}.$$

It is immediately checked that $\mathrm{Sp}_6(q^2).2$ contains a conjugate of z, while $\mathrm{Sp}_4(q^3).3$ does not, which produces (4). Assume next that $n \neq 12$. Then t is a prime. Let $s \in P_{2t}(q)$. Since s divides the order of M, we have that s divides $q^{2ri} - 1$, for some $i \in \{1, \ldots, n/(2r)\}$, or $s = r$. In the first case, by Lemma 2.3 (3), we deduce $2t \mid 2ri$ and hence $t \mid ri$. We show that $\gcd(t, r) = 1$. Assume the contrary. Then, since both r and t are prime, we have $t = r \mid n/2$, a contradiction. Thus we have $t \mid i$, which gives $t \leq n/4$, a contradiction. It follows that $s = r$. By (2.4), we then get $r = s > 2t > n/2$, contradicting the fact that r divides $n/2$.

Suppose next that M is of type $\mathrm{GU}_{n/2}(q).2$. Now, from [49] or from [23], $\mathrm{GU}_{n/2}(q)$ has elements of order $(q^t + 1)(q^{\frac{n}{2}-t} + 1)/\gcd(q^t + 1, q^{\frac{n}{2}-t} + 1)$ if and only if $n/2 - t$ is odd. Recall now that t is an odd prime when $n \neq 12$ and $t = 4$ when $n = 12$. Thus $n/2 - t$ is odd only when $n/2$ is even with $n \neq 12$. Thus part (2) holds.

The Subgroup M Lies in Class $\mathcal{C}_5$ This case is ruled out by Lemma 2.7 (3), because z is a strong $ppd(n, q; 2t)$-element and $n \geq 10$.

The Subgroup M Lies in Class $\mathcal{C}_8$ From [66, Section 4.8], q is even and M is of type $\mathrm{SO}_n^\pm(q)$. Now, $\mathrm{SO}_n^-(q)$ has no element[6] having order $\mathbf{o}(z) = \frac{(q^t+1)(q^{\frac{n}{2}-t}+1)}{\gcd(q^t+1, q^{\frac{n}{2}-t}+1)}$ and action type $2t \oplus (n - 2t)$. Thus M is of type $\mathrm{SO}_n^+(q)$ and part (3) holds.

The Subgroup M Lies in Class $\mathcal{S}$ From Example 2.6 a) of [53], we obtain

$$A_m \leq M \leq S_m \times \mathbf{Z}(\mathrm{Sp}_n(q)),$$

with $m \in \{n+1, n+2\}$.

Thus the symmetric group S_m contains an element having order

$$\frac{\mathbf{o}(z)}{\gcd(2, q-1)} = \frac{(q^t+1)(q^{\frac{n}{2}-t}+1)}{\gcd(q^t+1, q^{\frac{n}{2}-t}+1)\gcd(2, q-1)}.$$

We first deal with the case $n = 12$. Thus $t = 4$ and

$$\mathbf{o}(z) = \frac{(q^4+1)(q^2+1)}{\gcd(2, q-1)}.$$

Thus $\mathbf{o}(z) \geq q^6/2$. The maximal element order of S_{14} is 84 and hence $84 \geq q^6/4$. This yields $q = 2$. However, when $q = 2$, $\mathbf{o}(z) = 85$ and hence also this case does not arise. For the rest of our argument, we may suppose that $n \neq 12$. In particular, t is an odd prime and $t \nmid n/2$. Then (5.6) yields that $\gcd(q^t+1, q^{\frac{n}{2}-t}+1) \leq q+1$. Hence

$$\mathbf{o}(z) \geq \frac{q^{n/2}}{q+1}.$$

Let o be the maximal element order of an element in S_m. From [69] and [75, Theorem 2], we have

$$\log o \leq \sqrt{m \log m} \left(1 + \frac{\log(\log(m)) - a}{2 \log(m)}\right),$$

where $\log(m)$ denotes the logarithm of m to the base e and $a := 0.975$. Therefore

$$\log\left(\frac{q^{n/2}}{2(q+1)}\right) \leq \sqrt{m \log m} \left(1 + \frac{\log(\log(m)) - a}{2 \log(m)}\right). \quad (5.7)$$

---

[6] See the last paragraph of the proof of Lemma 5.5 or of Lemma 5.6.

5.2 Large Dimensional Symplectic Groups

This inequality holds true only when[7]

- $q = 4$ and $n \in \{10, 12\}$, or
- $q = 3$ and $n \in \{10, 12, 14, 16, 18\}$, or
- $q = 2$ and $n \leq 46$.

For these remaining cases, we have computed explicitly $o(z)$ and the order of the elements of S_{n+2} and we have verified that in no case $\frac{o(z)}{\gcd(2,q-1)}$ is the order of a permutation in S_{n+2}. □

Proposition 5.8 *For every $n \geq 10$, the weak normal covering number of $\mathrm{PSp}_n(q)$ is at least 3, unless q is even. When q is even, if H and K are maximal components of a weak normal 2-covering of $\mathrm{PSp}_n(q) = \mathrm{Sp}_n(q)$, then up to $\mathrm{Aut}(\mathrm{PSp}_n(q))$-conjugacy, we have $H \cong \mathrm{SO}_n^-(q)$ and $K \cong \mathrm{SO}_n^+(q)$. Such weak normal covering gives rise to a single normal covering up to conjugacy.*

Proof When $(n, q) \in \{(10, 2), (14, 2)\}$, the proof follows with a computer computation, therefore, for the rest of the proof we exclude the case $(n, q) \in \{(10, 2), (14, 2)\}$. Let $\tilde{H}$ be a component of a weak normal covering of $\mathrm{Sp}_n(q)$ containing a Singer cycle. Thus, $\tilde{H}$ is one of the groups in parts (1), (2) and (3) of Lemma 5.1. From [66, Table 3.5C], we see that in each of these three possibilities we have

$$\mathrm{Aut}(\mathrm{Sp}_n(q)) = \mathbf{N}_{\mathrm{Aut}(\mathrm{Sp}_n(q))}(\tilde{H}) \mathrm{Sp}_n(q).$$

Therefore, $\tilde{H}$ on its own cannot give rise to a weak normal covering of $\mathrm{Sp}_n(q)$. Thus $\gamma_w(\mathrm{Sp}_n(q)) \geq 2$. Let now μ be a weak normal 2-covering of $\mathrm{PSp}_n(q)$ with maximal components and with $\tilde{H} \in \tilde{\mu}$.

Now, the maximal subgroups of $\mathrm{Sp}_n(q)$ containing a Bertrand element are described in Lemma 5.7. By the above arguments, the corresponding weak normal 2-covering $\tilde{\mu}$ of $\mathrm{Sp}_n(q)$ is given by $\tilde{\mu} = \{\tilde{H}, \tilde{K}\}$, where $\tilde{K}$ must be in one of the possibilities described in Lemma 5.7.

Let $y \in \mathrm{Sp}_n(q)$ be of order

$$\frac{(q^{\frac{n}{2}-1} + 1)(q + 1)}{\gcd(q^{\frac{n}{2}-1} + 1, q + 1)}$$

---

[7] The function $q \mapsto q^{n/2}/(2(q + 1))$ is increasing in q and hence $\log(q^{n/2}/(2(q + 1))) \geq \log(2^{n/2}/6) \geq \log(2^{n/2-3}) = (n/2 - 3)\log(2)$.

The function $m \mapsto (1 + (\log(\log(m)) - a)/(2\log(m)))$ is bounded above by 2 and hence $\sqrt{m \log m}\left(1 + \frac{\log(\log(m))-a}{2\log(m)}\right) < 2\sqrt{m \log(m)} \leq 2\sqrt{(n+2)\log(n+2)}$.

It is now easy to verify that $(n/2 - 3)\log(2) > 2\sqrt{(n+2)\log(n+2)}$ when $n \geq 200$. We have implemented in a computer the inequality in (5.7) for dealing with small values of n.

and action of type $(n-2) \oplus 2$. Then V has V has only two proper y-invariant subspaces, say U and W, $V = U \perp W$, $\dim_{\mathbb{F}_q} U = n-2$ and $\dim_{\mathbb{F}_q} W = 2$. In particular, y has no $\mathrm{Aut}(\mathrm{Sp}_n(q))$-conjugate in $\mathrm{Sp}_{2t}(q) \perp \mathrm{Sp}_{n-2t}(q)$ because $2t \notin \{2, n-2\}$.

Observe that $P_{n-2}(q) \neq \emptyset$ and let $r \in P_{n-2}(q)$. If r divides the order of $\mathrm{Sp}_{n/k}(q^k).k$, then either r divides $q^{2ik} - 1$ for some $i \in \{1, \ldots, n/(2k)\}$ or $r = k$. However, both possibilities are impossible. Indeed, the first possibility yields $2ik \geq n-2$ and hence $i = n/(2k)$. However, when $i = n/(2k)$, we have $q^{2ik} - 1 = q^n - 1$ and $\gcd(q^n - 1, q^{n-2} - 1) = q^2 - 1$ by Lemma 2.2 (1). This yields that r divides $q^2 - 1$, which is a contradiction. Similarly, when $k = r$, we contradict (2.4).

The unitary group $\mathrm{GU}_{n/2}(q).2$ contains an element $\mathrm{Aut}(\mathrm{Sp}_n(q))$-conjugate to y (that is, of order $(q^{n/2-1} + 1)(q + 1)/\gcd(q^{n/2-1} + 1, q + 1)$ and of action type $2 \oplus (n-2)$) only when $n/2 - 1$ is odd, that is, $n/2$ is even.

When q is even and $\varepsilon \in \{-, +\}$, the orthogonal group $\mathrm{SO}_n^\varepsilon(q)$ contains an element $\mathrm{Aut}(\mathrm{Sp}_n(q))$-conjugate to y only when $\varepsilon = +$, because y lies in $\mathrm{SO}_{n-2}^-(q) \perp \mathrm{SO}_2^-(q)$.

Suppose now that $\tilde{H}$ is as in part (2) of Lemma 5.1, that is, $\tilde{H} = \mathrm{GU}_{n/2}(q).2$, with $nq/2$ odd. The condition on n and on q and Lemma 5.7 yield $\tilde{K} = \mathrm{Sp}_{2t}(q) \perp \mathrm{Sp}_{n-2t}(q)$. However, neither $\tilde{H}$ nor $\tilde{K}$ contain an $\mathrm{Aut}(\mathrm{Sp}_n(q))$-conjugate of y, which is a contradiction. Thus, $\tilde{H}$ cannot be as in part (2) of Lemma 5.1. Using this fact and using the fact that an $\mathrm{Aut}(\mathrm{Sp}_n(q))$-conjugate of y lies in $\tilde{H}$ or in $\tilde{K}$, by combining the possibilities for $\tilde{H}$ in Lemma 5.1 with the possibilities for $\tilde{K}$ in Lemma 5.7, we obtain that one of the following possibilities occurs

(1) $\tilde{H} \cong \mathrm{Sp}_{n/k}(q^k).k$ and $\tilde{K} \cong \mathrm{GU}_{n/2}(q).2$, with $n/2$ even, $n \neq 12$ and q odd,
(2) $\tilde{H} \cong \mathrm{Sp}_{n/k}(q^k).k$ and $\tilde{K} \cong \mathrm{SO}_n^+(q)$, with q even,
(3) $\tilde{H} \cong \mathrm{SO}_n^-(q)$ and $\tilde{K} \cong \mathrm{SO}_n^+(q)$, with q even.

We now consider each possibility in turn.

Case (3) is the Dye normal covering of $\mathrm{Sp}_n(q)$ appearing in the statement of our proposition and hence it requires no further comment. As usual it gives rise to a unique $\mathrm{Sp}_n(q)$-class of normal coverings. To exclude further cases, we use an element similar to y. Let $y' \in \mathrm{Sp}_n(q)$ be of order

$$\frac{(q^{\frac{n}{2}-2} + 1)(q^2 + 1)}{\gcd(q^{\frac{n}{2}-2} + 1, q^2 + 1)}$$

and action of type $(n-4) \oplus 4$. Then V has only two proper y'-invariant subspaces, say U and W, $V = U \perp W$, $\dim_{\mathbb{F}_q} U = n-4$ and $\dim_{\mathbb{F}_q} W = 4$.

Observe that $P_{n-4}(q) \neq \emptyset$ because we are excluding the case $(n, q) = (10, 2)$. Let $r \in P_{n-4}(q)$. If r divides the order of $\mathrm{Sp}_{n/k}(q^k).k$, then either r divides $q^{2ik} - 1$ for some $i \in \{1, \ldots, n/(2k)\}$ or $r = k$. However, both possibilities are impossible. Indeed, the first possibility yields $2ik \geq n-4 > n/2$ and hence $i = n/(2k)$.

5.2 Large Dimensional Symplectic Groups

However, when $i = n/(2k)$, we have $q^{2ik} - 1 = q^n - 1$ and $\gcd(q^n - 1, q^{n-4} - 1) = q^{\gcd(4,n)} - 1$ by Lemma 2.2 (1). This yields that r divides $q^4 - 1$, which is a contradiction because $r \in P_{n-4}(q)$. Similarly, when $k = r$, we contradict (2.4).

The unitary group $\mathrm{GU}_{n/2}(q).2$ contains an element $\mathrm{Aut}(\mathrm{Sp}_n(q))$-conjugate to y' (that is, of order $(q^{n/2-2} + 1)(q^2 + 1)/\gcd(q^{n/2-2} + 1, q^2 + 1)$ and of action type $(n-4) \oplus 4$) only when $n/2 - 2$ is odd, that is, $n/2$ is odd.

Using our considerations on y', we have excluded Case (1) and hence we are left to discuss Case (2).

Let $y'' \in \mathrm{Sp}_n(q)$ be of order

$$\frac{(q^{\frac{n}{2}-1} - 1)(q + 1)}{\gcd(q^{\frac{n}{2}-1} - 1, q + 1)}$$

and action of type $(n/2 - 1) \oplus (n/2 - 1) \oplus 2$.

Here we aim to prove that y'' has no $\mathrm{Aut}(\mathrm{Sp}_n(q))$-conjugate in $\mathrm{Sp}_{n/k}(q^k).k$. We argue by contradiction and we suppose, without loss of generality, that $y'' \in \mathrm{Sp}_{n/k}(q^k).k$. Observe that $P_{n/2-1}(q) \neq \emptyset$, because we are excluding the case $(n, q) = (14, 2)$. Let $r \in P_{n/2-1}(q)$. If r divides the order of $\mathrm{Sp}_{n/k}(q^k).k$, then either r divides $q^{2ik} - 1$ for some $i \in \{1, \ldots, n/(2k)\}$ or $r = k$. The first possibility yields $r \mid \gcd(q^{2ik} - 1, q^{n/2-1} - 1) = q^{\gcd(2ik, n/2-1)} - 1$, by Lemma 2.2. Since $r \in P_{n/2-1}(q)$, we must have $\gcd(n/2 - 1, 2ik) = n/2 - 1$ and hence

$$2ik = \ell\left(\frac{n}{2} - 1\right),$$

for some $\ell \in \mathbb{N}$ with $\ell \neq 0$. As $2ik \leq n$, we obtain $\ell \in \{1, 2\}$. When $\ell = 1$, we get $2ik = n/2 - 1$ and k divides $n/2 - 1$. As k divides also $n/2$, we have $k \mid \gcd(n/2 - 1, n/2) = 1$, which is a contradiction. Analogously, when $\ell = 2$, we get $2ik = n - 2$ and $i = n/(2k) - 1/k$. Since $n/2k$ is an integer, we obtain another contradiction. Similarly, when $k = r$, (2.4) yields $k = r \geq n/2 - 1 + 1 = n/2$ and hence $r = k = n/2$. Summing up, $y'' \in \mathrm{Sp}_2(q^{n/2}).(n/2)$ and hence

$$q^{\frac{n}{2}-1} - 1 = \gcd\left(q^{\frac{n}{2}-1} - 1, |\mathrm{Sp}_2(q^{n/2}).(n/2)|\right)$$

$$\leq \gcd(q^{\frac{n}{2}-1} - 1, |\mathrm{Sp}_2(q^{n/2})|) \cdot \frac{n}{2}$$

$$= \gcd(q^{\frac{n}{2}-1} - 1, q^n - 1) \cdot \frac{n}{2} \leq (q^{\gcd(n/2-1,n)} - 1)\frac{n}{2}$$

$$= (q^2 - 1)\frac{n}{2}.$$

However, this inequality is never satisfied.

When q is even and $\varepsilon \in \{-, +\}$, the orthogonal group $\mathrm{SO}_n^\varepsilon(q)$ contains an element $\mathrm{Aut}(\mathrm{Sp}_n(q))$-conjugate to y'' only when $\varepsilon = -$, because y'' lies in $\mathrm{SO}_{n-2}^+(q) \perp \mathrm{SO}_2^-(q)$. Using our considerations on y'', we finally exclude Case (2). □

Now, the veracity of Table 1.6 follows from the results in this chapter.

Chapter 6
Odd Dimensional Orthogonal Groups

In this section we are concerned with $P\Omega_n(q)$ with nq odd and $n \geq 7$. Recall that $P\Omega_n(q) = \Omega_n(q)$. Note that the case $n = 3$ is considered in Chap. 3 because $\Omega_3(q) \cong \mathrm{PSL}_2(q)$, the case $n = 5$ is considered in Chap. 5 because $\Omega_5(q) \cong \mathrm{PSp}_4(q)$, the case q even is considered also in Chap. 5 because $\Omega_n(q) \cong \mathrm{PSp}_{n-1}(q)$.

Lemma 6.1 ([74, Theorem 1.1]) *Let M be a maximal subgroup of $\Omega_n(q)$ containing a semisimple element having order $(q^{\frac{n-1}{2}} + 1)/2$ and action type $1 \oplus (n-1)$. Then, one of the following holds*

(1) $M \cong \Omega_{n-1}^{-}(q).2 \in \mathcal{C}_1$,
(2) $M \cong S_9 \in \mathcal{S}$, $n = 7$ and $q = 3$.

In earlier sections, in our analysis of linear, unitary and symplectic groups, we have used mostly semisimple elements. For the orthogonal groups $\Omega_n(q)$, it is instead convenient to make a massive use of unipotent elements. Therefore, before proving our main result for $\Omega_n(q)$, we collect here some basic remarks about those elements that we need in the sequel.

Recall that $u \in \mathrm{GL}_n(q)$ is **regular unipotent**, if $u - I_n$ has rank $n - 1$, where I_n denotes the $n \times n$ identity matrix. The odd dimensional orthogonal group $\Omega_n(q)$ contains regular unipotent elements. Indeed, the existence of regular unipotent elements in $\mathrm{SO}_n(q)$ follows from [27, Proposition 5.7.1]; however, since $|\mathrm{SO}_n(q) : \Omega_n(q)| = 2$ (except when $n = 1$) and since q is odd, we deduce that the regular unipotent elements of $\mathrm{SO}_n(q)$ all lie in $\Omega_n(q)$.

Let u be a regular unipotent element. Clearly, the Jordan form of u consists of a unique Jordan block, that is, u is conjugate via an element of $\mathrm{GL}_n(q)$ to the matrix

$$\begin{pmatrix} 1 & 0 & \cdots & & & \cdots & 0 \\ 1 & 1 & 0 & & & & \vdots \\ 0 & 1 & 1 & 0 & & & \\ \vdots & \ddots & \ddots & \ddots & & & \\ \vdots & & & \ddots & \ddots & \ddots & \vdots \\ 0 & \cdots & & & 0 & 1 & 1 & 0 \\ 0 & \cdots & & & \cdots & 0 & 1 & 1 \end{pmatrix}.$$

Using this representative of the conjugacy class of u in $\mathrm{GL}_n(q)$, an elementary computation shows that, for every $i \in \{0,\ldots,n\}$, there exists a unique $\mathbb{F}_q\langle u\rangle$-submodule V_i of $V = \mathbb{F}_q^n$ with $\dim_{\mathbb{F}_q}(V_i) = i$. Indeed, if we let $e_1,\ldots,e_n$ be the standard basis of V, we have $V_0 = 0$ and $V_i = \langle e_1,\ldots,e_i\rangle$ for $i \in \{1,\ldots,n\}$. In particular, the $n+1$ u-invariant subspaces $V_0,\ldots,V_n$ of V form the flag

$$0 = V_0 < V_1 < V_2 < \cdots < V_{n-1} < V_n = V$$

and u has a unique u-invariant subspace of dimension 1 given by V_1.

Suppose now that u preserves the non-degenerate quadratic form Q on V. As V_1 is u-invariant, so is $V_1^\perp$. Since Q is non-degenerate, $\dim_{\mathbb{F}_q}(V_1^\perp) = \dim_{\mathbb{F}_q}(V) - \dim_{\mathbb{F}_q}(V_1) = n - 1$. Thus $V_1^\perp = V_{n-1}$ and hence $V_1 \le V_1^\perp$. Therefore, V_1 is totally singular.

6.1 Small Odd Dimensional Orthogonal Groups

We start our analysis with small odd dimensional orthogonal groups. For the subgroup structure of $\Omega_7(q)$ and $\Omega_9(q)$ we use [9].

Lemma 6.2 *The weak normal covering number of $\Omega_7(q)$ is at least* 3.

Proof When $q = 3$, the result follows with a computation with the computer algebra system magma [6]. Therefore, for the rest of the argument, we suppose $q \ge 5$.

Let μ be a weak normal covering of $\Omega_7(q)$ of minimum cardinality and with maximal components. Let $H \in \mu$ with H containing a semisimple element of order $(q^3+1)/2$ and action type $1 \oplus 6$. From Lemma 6.1, we have that

$$H \cong \Omega_6^-(q).2.$$

6.1 Small Odd Dimensional Orthogonal Groups

A case-by-case analysis on the maximal subgroups of $\Omega_7(q)$ in [9, Tables 8.39, 8.40] reveals that there are two $\Omega_7(q)$-conjugacy classes of maximal subgroups containing elements of order $q^3 - 1$. Namely,

(1) $E_q^{6+6} : \frac{1}{2}\mathrm{GL}_3(q) \in \mathcal{C}_1$,
(2) $\Omega_6^+(q).2 \in \mathcal{C}_1$.

In particular, since H is not in this list, we deduce that there exists a further component $K \in \mu$ and that K is in one of the above two possibilities.

Now, let $u \in \Omega_3(q)$ be a regular unipotent element, let $s \in \Omega_4^-(q)$ be a semisimple element of order $(q^2 + 1)/2$ and having type 4 on $\mathbb{F}_q^4$ and let $z = u \oplus s$. By construction, $z \in \Omega_7(q)$ with respect to a suitable non-degenerate quadratic form. Since u has order p, we have $\mathbf{o}(z) = p(q^2 + 1)/2$.

As we observed above, the only 1-dimensional subspace of $\mathbb{F}_q^3$ left invariant by u is totally singular. Therefore, the only 1-dimensional subspace of V left invariant by z is totally singular and hence z is not $\mathrm{Aut}(\Omega_7(q))$-conjugate to elements in $\Omega_6^+(q).2$ or $\Omega_6^-(q).2$, because these groups stabilize a 1-dimensional non-singular subspace. However, no element in $E_q^{6+6} : \frac{1}{2}\mathrm{GL}_3(q)$ has order divisible by $(q^2 + 1)/2$: this can be seen using Lemma 2.2 and using the fact that $q \geq 5$ is odd. Thus $\gamma_w(\Omega_7(q)) \geq 3$. □

Lemma 6.3 *The weak normal covering number of $\Omega_9(q)$ is at least 3.*

Proof Let μ be a weak normal covering of $\Omega_9(q)$ of minimum cardinality and with maximal components. Let $H \in \mu$ with H containing a semisimple element of order $(q^4 + 1)/2$ and action type $1 \oplus 8$. From Lemma 6.1, we have that

$$H \cong \Omega_8^-(q).2.$$

Now, let $u \in \Omega_3(q)$ be a regular unipotent element, let $s \in \Omega_6^-(q)$ be a semisimple element of order $(q^3 + 1)/2$ and having type 6 on $\mathbb{F}_q^6$ and let $y = u \oplus s \in \mathrm{GL}_9(q)$. By construction, $y \in \Omega_9(q)$ with respect to a suitable non-degenerate quadratic form. Since u has order p, we have $\mathbf{o}(y) = p(q^3 + 1)/2$.

A case-by-case analysis on the maximal subgroups of $\Omega_9(q)$ in [9, Tables 8.58, 8.59] reveals that there are two $\Omega_9(q)$-conjugacy classes of maximal subgroups containing a conjugate of y. Namely,

(1) $E_q^7 : \left(\frac{q-1}{2} \times \Omega_7(q)\right).2 \in \mathcal{C}_1$,
(2) $(\Omega_3(q) \times \Omega_6^-(q)).2^2 \in \mathcal{C}_1$.

In particular, since H is not in this list, there exists a further component $K \in \mu$ such that K is in one of the above two possibilities. Suppose $\mu = \{H, K\}$.

Let u' be a regular unipotent element of $\Omega_9(q)$. As we observed above, the only 1-dimensional subspace of $V = \mathbb{F}_q^9$ left invariant by u' is totally singular. Every unipotent element v in $(\Omega_3(q) \times \Omega_6^-(q)).2^2$ lies in $\Omega_3(q) \times \Omega_6^-(q)$ and hence satisfies $\dim_{\mathbb{F}_q} C_V(v) \geq 2$. In particular, u' has no $\text{Aut}(\Omega_9(q))$-conjugate in H (because u' does not fix a non-degenerate vector) or in $(\Omega_3(q) \times \Omega_6^-(q)).2^2$. We deduce that

$$K \cong E_q^7 : \left(\frac{q-1}{2} \times \Omega_7(q)\right).2.$$

Now, let z be a semisimple element of $\Omega_9(q)$ belonging to the maximal subgroup $\Omega_8^+(q).2$ and having type $4 \oplus 4 \oplus 1$. This element can be explicitly constructed by taking the quadratic form Q preserved by $\Omega_9(q)$ and letting $z \in \Omega_9(q)$ be a semisimple element fixing a vector $v \in V$ with $Q(v)$ not a square in $\mathbb{F}_q$ and with z inducing on $\langle v \rangle^\perp$ a semisimple matrix having type $4 \oplus 4$. Now, z lies in no conjugate of $H \cong \Omega_8^-(q).2$ because z does not fix any non-zero vector w with $Q(w)$ a square in $\mathbb{F}_q$. As we are assuming that $\mu = \{H, K\}$ is a weak normal covering of $\Omega_9(q)$, z is $\text{Aut}(\Omega_9(q))$-conjugate to an element of K. Since z is semisimple, z is $\text{Aut}(\Omega_9(q))$-conjugate to a semisimple element in $(\frac{q-1}{2} \times \Omega_7(q)).2$. However, since z has type $4 \oplus 4 \oplus 1$, z preserves no 7-dimensional subspace of V. $\square$

6.2 Large Odd Dimensional Orthogonal Groups

In this section, we deal with large odd dimensional orthogonal groups $\Omega_n(q)$ with $n \geq 11$. We consider the Bertrand number t for the integer $\frac{n-1}{2} \geq 5$ (see Sect. 2.10). Thus for $n \geq 17$ we have $\frac{n-1}{4} < t \leq \frac{n-1}{2} - 3$ and t is an odd prime; for $n = 15$, we have $t = 5$; for $n = 13$, we have $t = 4$; for $n = 11$ we have $t = 3$. In particular t is always an odd prime apart the case $n = 13$. Moreover, for every $n \geq 11$, we have $\frac{n-1}{4} < t \leq \frac{n-1}{2} - 2$ and $t \nmid \frac{n-1}{2}$. Thus, from Lemma 2.2 (2), we have

$$\gcd(q^t+1, q^{\frac{n-1}{2}-t}+1) = \begin{cases} q+1 & \text{if } \frac{n-1}{2} \text{ is even and } n \neq 13, \\ 2 & \text{if } \frac{n-1}{2} \text{ is odd or } n = 13. \end{cases} \quad (6.1)$$

We consider a Bertrand element z of $\Omega_n(q)$ such that

- $o(z) = \frac{(q^t+1)(q^{\frac{n-1}{2}-t}+1)}{\gcd(q^t+1, q^{\frac{n-1}{2}-t}+1)}$ (see Table 2.2),
- the action of z on V is of type $1 \oplus 2t \oplus (n - 2t - 1)$ and we write $V = U_1 \perp U_2 \perp U_3$, where $\dim_{\mathbb{F}_q} U_1 = 1$, $\dim_{\mathbb{F}_q} U_2 = 2t$, $\dim_{\mathbb{F}_q} U_3 = n - 2t - 1$ and U_1, U_2, U_3 are irreducible $\mathbb{F}_q \langle z \rangle$-modules,
- z induces a matrix of order $q^t + 1$ on U_2 and a matrix of order $q^{\frac{n-1}{2}-t}+1$ on U_3.

6.2 Large Odd Dimensional Orthogonal Groups

The existence of Bertrand elements is easily proved. Consider a Singer cycle $s_{2t} \in$ $\mathrm{SO}_{2t}^-(q)$ and a Singer cycle $s_{n-2t+1} \in \mathrm{SO}_{n-2t-1}^-(q)$. Then, by Proposition 2.8, we have $s_{2t} \oplus s_{n-2t+1} \in \Omega_{n-1}^+(q)$ so that $z := I_1 \oplus s_{2t} \oplus s_{n-2t+1} \in \Omega_n(q)$ is a Bertrand element of $\Omega_n(q)$.

Observe that, since q is odd and both $2t$ and $n - 2t - 1$ are at least 4, the sets $P_{2t}(q)$ and $P_{n-2t-1}(q)$ are both non-empty. Moreover, since $2t \neq n - 2t - 1$, Lemma 2.3 (2) implies that $P_{2t}(q) \cap P_{n-2t-1}(q) = \emptyset$. In particular, for every $a \in P_{2t}(q)$ and $b \in P_{n-2t-1}(q)$, the product ab divides $\mathbf{o}(z)$. In particular

$$z \text{ is a strong } ppd(n, q; 2t)\text{-element.}$$

Lemma 6.4 *If M is a maximal subgroup of $\Omega_n(q)$ with $n \geq 11$ containing a Bertrand element, then one of the following holds*

(1) *M is of type $\mathrm{O}_{2t+1}(q) \perp \mathrm{O}_{n-2t-1}^-(q) \in \mathcal{C}_1$,*
(2) *M is of type $\mathrm{O}_{n-2t}(q) \perp \mathrm{O}_{2t}^-(q) \in \mathcal{C}_1$,*
(3) *M is of type $\mathrm{O}_{n-1}^+(q) \in \mathcal{C}_1$.*

Proof Let z be a Bertrand element of $\Omega_n(q)$ with $n \geq 11$. By Theorem 2.6, Lemma 2.7 (3) and [66, Table 3.5D], M belongs to one of the Aschbacher classes $\mathcal{C}_i, i \in \{1, 2, 3\}$ or to $\mathcal{S}$ and is described in Example 2.6 a) of [53].

The Subgroup M Lies in Class $\mathcal{C}_1$ Suppose that M is the stabilizer of a totally singular subspace of V of dimension m with $1 \leq m \leq (n-1)/2$, that is, M is a parabolic subgroup P_m. The element z acts irreducibly on U_1, U_2 and U_3 and U_1, U_2, U_3 are non-degenerate of dimension $1, 2t$ and $n - 2t - 1$ respectively. Hence U_1, U_2, U_3 are pairwise non-isomorphic and they are the only irreducible $\mathbb{F}_q\langle z\rangle$-submodules of V. If $z \in M$, then z fixes a non-trivial totally singular subspace U and hence $U_i \leq U$, for some i. However, this contradicts the fact that U_i is non-degenerate.

Suppose now that M is the stabilizer of a non-degenerate subspace U of V of dimension $2k + 1$, that is, M is of type $\mathrm{O}_{2k+1}(q) \perp \mathrm{O}_{n-2k-1}^\epsilon(q)$, with $\epsilon \in \{+, -\}$. We refer to [66, Proposition 4.1.6] for the precise structure of M.

Since U must be the sum of irreducible $\mathbb{F}_q\langle z\rangle$-submodules of V, and the only possible dimensions for an irreducible $\mathbb{F}_q\langle z\rangle$-submodule of V are $1, 2t$ and $n-2t-1$, then the only possibilities for the dimension of U are $2k + 1 = 1$, $2k + 1 = 2t + 1$, $2k + 1 = n - 2t$, that is, $k \in \{0, t, \frac{n-2t-1}{2}\}$. Thus, in order to reach (1)–(3), we just need to show that when $k = 0$ the only possibility is $\epsilon = +$, while for the other two cases the only possibility is $\epsilon = -$.

We discuss the three cases. Suppose first that $k = 0$. Now, $z \in \mathrm{O}_1(q) \perp \mathrm{O}_{n-1}^\epsilon(q)$. The $(n-1)$-dimensional subspace of V fixed by z is $U_2 \oplus U_3$. Observe that z restricted to U_2 is the Singer cycle s_{2t} and z restricted to U_3 is the Singer cycle s_{n-2t-1}. In particular, the quadratic form restricted to U_2 and to U_3 must be of minus type. Therefore, the quadratic form induced on $U_2 \oplus U_3$ is of plus type and hence $\epsilon = +$. Suppose next that $k = t$. Now, $z \in \mathrm{O}_{2t+1}(q) \perp \mathrm{O}_{n-2t-1}^\epsilon(q)$. The

$(n - 2t - 1)$-dimensional subspace of V fixed by z is U_3. Observe that z restricted to U_3 is the Singer cycle s_{n-2t-1}. In particular, the quadratic form restricted to U_3 must be of minus type and hence $\epsilon = -$. The argument when $t = (n - 2t - 1)/2$ is similar to the previous case and thus omitted.

The Subgroup M Lies in Class C_2 These are the groups in the Example 2.3 of [53]. Thus $M \leq \mathrm{GL}_1(q) \mathrm{wr} S_n$. It follows that, for every $a \in P_{2t}(q)$ and $b \in P_{n-2t-1}(q)$, the product ab divides the order of an element in S_n. Since a and b are distinct primes, that requires $n \geq a + b$ and hence, by (2.4), $n \geq 2t + 1 + n - 2t = n + 1$, which is a contradiction.

The Subgroup M Lies in Class C_3 These groups are described in [53, Example 2.4]. Recall that z is a $ppd(n, q; 2t)$ element. Since $n \neq 2t + 1$, we consider only the case b) of Example 2.4 in [53]. Hence $M \leq \mathrm{GL}_{n/b}(q^b).b$, where b is a divisor of $\gcd(n, 2t)$ with $b \geq 2$. As n is odd, b is odd and hence $b = t \mid n$ so that $n = kt$, for some positive integer k. Since $(n - 1)/4 < t < (n - 1)/2$, we have that $k = 3$ divides n and $b = t = n/3$. Thus $n \geq 15$, so that t is an odd prime and, by Proposition 4.3.17 in [66], we have

$$M \cong \Omega_3(q^t).t$$

and M has size $\frac{q^t(q^{2t}-1)t}{2}$. Assume, by contradiction, that $z \in M$. Then $o(z) \mid |M|$ and thus

$$\frac{q^{\frac{n-1}{2}-t} + 1}{\gcd(q^t + 1, q^{\frac{n-1}{2}-t} + 1)} \mid \frac{q^t - 1}{2} t. \tag{6.2}$$

By Lemma 2.2 (3), recalling that t is odd, we have that

$$\gcd(q^{\frac{n-1}{2}-t} + 1, q^t - 1) = 2.$$

Moreover, by (6.1), we have that $2 \mid \gcd(q^t + 1, q^{\frac{n-1}{2}-t} + 1)$. Hence, by (6.2), we deduce that

$$\frac{q^{\frac{n-1}{2}-t} + 1}{\gcd(q^t + 1, q^{\frac{n-1}{2}-t} + 1)} \mid t. \tag{6.3}$$

Recalling that $t = n/3$ and that $\gcd(q^t + 1, q^{\frac{n-1}{2}-t} + 1) \leq q + 1$, it follows that

$$\frac{q^{\frac{n-3}{6}} + 1}{q + 1} \leq \frac{n}{3}. \tag{6.4}$$

6.2 Large Odd Dimensional Orthogonal Groups

Now if $n \geq 27$ we have

$$\frac{q^{\frac{n-3}{6}}+1}{q+1} \geq q^{\frac{n-3}{6}-2}(q-1) \geq 2q^{\frac{n-15}{6}} \geq 2 \cdot 3^{\frac{n-15}{6}} > \frac{n}{3},$$

which contradicts (6.4).

It remains to consider the cases $n \in \{15, 21\}$. Let $n = 15$. Then $t = 5$ and (6.3) becomes

$$\frac{q^2+1}{2} \mid 5,$$

which gives $q = 3$. But then we have $\mathbf{o}(z) = 5(3^5+1)$ and $M \cong \Omega_3(3^5).5$. Since $\Omega_3(3^5)$ has size $\frac{3^5(3^{10}-1)}{2}$, we have that $z \notin \Omega_3(3^5)$ and that a power of z is a generator[1] of $\mathrm{Gal}(\mathbb{F}_{3^5}/\mathbb{F}_3) = \langle \alpha \rangle$. Now $\mathbf{C}_M(\alpha) \cong \Omega_3(3).5$ and since z centralizes α we reach the contradiction $5(3^5+1) \mid 15(3^2-1)$.

Let finally $n = 21$. Then $t = 7$ and (6.3) becomes

$$\frac{q^3+1}{q+1} \mid 7,$$

which gives $q = 3$. But then we have $\mathbf{o}(z) = 7(3^7+1)$ and $M \cong \Omega_3(3^7).7$. Since $\Omega_3(3^7)$ has size $\frac{3^7(3^{14}-1)}{2}$, we have that $z \notin \Omega_3(3^7)$ and that a power of z is a generator of $\mathrm{Gal}(\mathbb{F}_{3^7}/\mathbb{F}_3) = \langle \alpha \rangle$. Now $\mathbf{C}_M(\alpha) \cong \Omega_3(3).7$ and since z centralizes α we reach the contradiction $7(3^7+1) \mid 21(3^2-1)$.

The Subgroup M Lies in Class $\mathcal{S}$ Let p be the characteristic of $\mathbb{F}_q$. The maximal subgroups M in Example 2.6 a) satisfy $M \leq S_m$, with $m \in \{n+1, n+2\}$.

From (6.1) and from a computation, we have

$$\mathbf{o}(z) > \frac{q^{\frac{n-1}{2}}}{q+1}.$$

Observe that $\mathbf{o}(z)$ is the order of an element in S_m. Arguing as in Lemma 5.7, we obtain

$$\log\left(\frac{q^{\frac{n-1}{2}}}{q+1}\right) \leq \sqrt{m \log m}\left(1 + \frac{\log(\log(m)) - 0.975}{2\log(m)}\right). \tag{6.5}$$

---

[1] This fact can be demonstrated using either the Lang–Steinberg theorem or, in a more elementary manner, by noting that 5 is coprime to the order of $\Omega_3(3^5)$, implying that all cyclic subgroups of order 5 within $\Omega_3(3^5).5$ are conjugate to the group of field automorphisms.

This inequality holds true only when[2] $n \in \{11, 13, 15, 17, 19\}$ and $q = 3$. We have checked those remaining cases with a computer and in no case S_{n+2} contains an element having order $o(z)$. □

Proposition 6.5 *For every $n \geq 11$, the weak normal covering number of $\Omega_n(q)$ is at least 3.*

Proof Assume, by contradiction, that H and K are maximal components of a weak normal 2-covering of $\Omega_n(q)$. By Lemma 6.1, we have $H \cong \Omega^-_{n-1}(q).2$ and K is given by one of the possibilities in Lemma 6.4.

Suppose first that K is of type $O_{2t+1}(q) \perp O^-_{n-2t-1}(q)$ or $O_{n-2t}(q) \perp O^-_{2t}(q)$. Without loss of generality, we may suppose that the orthogonal form for $\Omega_n(q)$ is given by the matrix

$$J := \begin{pmatrix} 0_{\frac{n-1}{2}} & I_{\frac{n-1}{2}} & 0 \\ I_{\frac{n-1}{2}} & 0_{\frac{n-1}{2}} & 0 \\ 0 & 0 & 1 \end{pmatrix},$$

where $0_{\frac{n-1}{2}}$ and $I_{\frac{n-1}{2}}$ are the $(n-1)/2 \times (n-1)/2$ zero and identity matrix respectively. We let $e_1, \ldots, e_n$ be the canonical basis of $V = \mathbb{F}_q^n$. Now, given a Singer cycle s for $\mathrm{GL}_{(n-1)/2}(q)$, we get

$$y' := \begin{pmatrix} s & 0 & 0 \\ 0 & (s^{-1})^T & 0 \\ 0 & 0 & 1 \end{pmatrix} \in \mathrm{SO}_n(q).$$

Set $y := y'^2$ and observe that y is a semisimple element in $\Omega_n(q)$. We claim that

(†) $U_1 := \langle e_1, \ldots, e_{(n-1)/2} \rangle$, $U_2 := \langle e_{(n+1)/2}, \ldots, e_{n-1} \rangle$ and $U_3 := \langle e_n \rangle$ are the only irreducible $\mathbb{F}_q \langle y \rangle$-submodules of V.

Indeed, s^{q-1} is a Singer cycle for $\mathrm{SL}_{(n-1)/2}(q)$ so that it acts irreducibly on U_1. Since the order of $((s^{-1})^T)^{q-1}$ is equal to the order of s^{q-1}, by Proposition 2.5, we have that also $((s^{-1})^T)^{q-1}$ is a Singer cycle for $\mathrm{SL}_{(n-1)/2}(q)$ so that it acts irreducibly on U_2.

Since q is odd, we have that $\langle s^2 \rangle \geq \langle s^{q-1} \rangle$ and thus s^2 acts irreducibly on U_1 and, similarly, $(s^{-2})^T$ acts irreducibly on U_2. As $\dim_{\mathbb{F}_q}(U_3) = 1$, it follows that U_1, U_2, U_3 are irreducible $\mathbb{F}_q \langle y \rangle$-submodules of V. It remains to prove the

---

[2] The argument here is similar to the proof of Lemma 5.7. The function $q \mapsto q^{(n-1)/2}/(q+1)$ is increasing in q and hence $\log(q^{(n-1)/2}/(q+1)) \geq \log(3^{(n-1)/2}/4) \geq \log(3^{(n-1)/2-2}) = ((n-5)/2) \log(3)$. Moreover, as in the proof of Lemma 5.7, we have $\sqrt{m \log m} \left(1 + \frac{\log(\log(m))-a}{2 \log(m)}\right) \leq 2\sqrt{(n+2) \log(n+2)}$.

It is now easy to verify that $((n-5)/2) \log(3) > 2\sqrt{(n+2) \log(n+2)}$ when $n \geq 70$. We have implemented in a computer the inequality in (6.5) for dealing with small values of n.

6.2 Large Odd Dimensional Orthogonal Groups

uniqueness. The uniqueness is clear when $U_1 \not\cong U_2$. Therefore for the rest of the proof of the claim we assume $U_1 \cong U_2$. Let $\varphi : U_1 \to U_2$ be an $\mathbb{F}_q\langle y\rangle$-isomorphism, that is, $\varphi(uz) = \varphi(u)z$ holds for all $u \in U_1$ and $z \in \langle y\rangle$. Represent now φ, with respect to the basis $(e_1, \ldots, e_{(n-1)/2})$ of U_1 and to the basis $(e_{(n+1)/2}, \ldots, e_{n-1})$ of U_2, via the matrix $B \in \mathrm{GL}_{(n-1)/2}(q)$. Then we have

$$B^{-1}s^2 B = (s^{-2})^T. \tag{6.6}$$

We know that, up to conjugacy, the Singer cycle $s \in \mathrm{GL}_{(n-1)/2}(q)$ is given by π_a where a is a generator of $\mathbb{F}^*_{q^{(n-1)/2}}$. In particular, the eigenvalues of s in the field $\mathbb{F}_{q^{(n-1)/2}}$ form the set $\{a, a^q, \ldots, a^{q^{\frac{n-1}{2}-1}}\}$. Thus the eigenvalues of s^2 in $\mathbb{F}_{q^{(n-1)/2}}$ form the set

$$\{a^2, a^{2q}, \ldots, a^{2q^{\frac{n-1}{2}-1}}\},$$

and the eigenvalues of $(s^{-2})^T$ in $\mathbb{F}_{q^{(n-1)/2}}$ form the set

$$\{a^{-2}, a^{-2q}, \ldots, a^{-2q^{\frac{n-1}{2}-1}}\}.$$

Now, (6.6) implies that

$$\{a^2, a^{2q}, \ldots, a^{2q^{\frac{n-1}{2}-1}}\} = \{a^{-2}, a^{-2q}, \ldots, a^{-2q^{\frac{n-1}{2}-1}}\}.$$

Therefore, there exists $i \in \{1, \ldots, (n-3)/2\}$ such that $a^2 = a^{-2q^i}$, that is, $a^{2(q^i+1)} = 1$. This implies that $q^{(n-1)/2} - 1$ divides $2(q^i + 1)$ so that, by Lemma 2.2, we deduce

$$q^{(n-1)/2} - 1 = \gcd(q^{(n-1)/2} - 1, 2(q^i + 1)) \leq 2(q^{\gcd(i,(n-1)/2)} + 1).$$

Since $i < (n-1)/2$, we have that $\gcd(i, (n-1)/2) \leq (n-1)/4$ and thus we get

$$q^{(n-1)/2} - 1 \leq 2(q^{(n-1)/4} + 1),$$

which implies $q^{(n-1)/4} \leq 3$. This is impossible because $q^{(n-1)/4} \geq 3^{10/4}$. This concludes the proof of our claim (†).

It is readily seen that the quadratic form induced by J on the subspace of $V = \mathbb{F}_q^n$ spanned by $e_1, \ldots, e_{n-1}$ has Witt defect 0. We have proven above that U_1, U_2, U_3 are the only irreducible $\mathbb{F}_q\langle y\rangle$-submodules of the semisimple module V. Therefore, the only proper $\mathbb{F}_q\langle y\rangle$-submodules of the semisimple module V are

$$U_1, U_2, U_3, U_1 \oplus U_2, U_1 \oplus U_3, U_2 \oplus U_3.$$

Recalling that U_1 and U_2 are totally isotropic of dimension $(n - 1)/2 > 1$, we deduce that the only proper non-degenerate $\mathbb{F}_q\langle y \rangle$-submodules of the semisimple module V are $U_1 \oplus U_2$ having dimension $n - 1$ and U_3 having dimension 1. What is more, the quadratic form induced on the only $(n - 1)$-dimensional y-invariant subspace of V, that is on $U_1 \oplus U_2$, is of Witt defect 0. From this, it follows that y is not $\mathrm{Aut}(\Omega_n(q))$-conjugate to an element in $H = \Omega_{n-1}^-(q).2$. Furthermore, recalling $t \leq \frac{n-1}{2} - 2$, we immediately check that $1, n - 1 \notin \{2t + 1, n - 2t - 1, 2t, n - 2t\}$ and thus y is not $\mathrm{Aut}(\Omega_n(q))$-conjugate to an element in K.

Finally, suppose that $K \cong \Omega_n^+(q).2$. Here we argue using regular unipotent elements. Let u be a regular unipotent element of $\Omega_n(q)$. Recall that the only 1-dimensional subspace of $V = \mathbb{F}_q^n$ left invariant by u is totally singular. Since the elements in $H \cong \Omega_n^-(q).2$ and in $K \cong \Omega_n^+(q).2$ all fix some non-degenerate 1-dimensional subspace of V, we deduce that u has no $\mathrm{Aut}(\Omega_n(q))$-conjugate in H or in K. □

Summing up, when $n \geq 7$, $\mathrm{P}\Omega_n(q) = \Omega_n(q)$ has weak normal covering number at least 3.

6.3 An Auxiliary Result

We include here a rather technical lemma which will only be used in the proof of Theorem 11.1 for even characteristic symplectic groups.

Lemma 6.6 *Let $n \geq 4$ be even and let q be a power of 2. Then $\mathrm{Sp}_n(q)$ contains two regular unipotent elements u_+ and u_- such that*

(1) *u_+ and u_- lie in distinct $\mathrm{Aut}(\mathrm{Sp}_n(q))$-conjugacy classes,*
(2) *u_+ preserves a quadratic form of Witt defect 0, u_- preserves a quadratic form of Witt defect 1, thus $u_+ \in \mathrm{SO}_n^+(q)$ and $u_- \in \mathrm{SO}_n^-(q)$ with respect to these quadratic forms,*
(3) *u_+ has no $\mathrm{Aut}(\mathrm{Sp}_n(q))$-conjugate in $\mathrm{SO}_n^-(q)$ and u_- has no $\mathrm{Aut}(\mathrm{Sp}_n(q))$-conjugate in $\mathrm{SO}_n^+(q)$,*
(4) *$u_+ \in \mathrm{SO}_n^+(q) \setminus \Omega_n^+(q)$ and $u_- \in \mathrm{SO}_n^-(q) \setminus \Omega_n^-(q)$.*

Proof Let $q = 2^f$, let $m := n/2$ and let

$$J := \begin{pmatrix} & & & & 1 \\ & & & 1 & \\ & & \iddots & & \\ & 1 & & & \\ 1 & & & & \end{pmatrix} \in \mathrm{GL}_m(q),$$

6.3 An Auxiliary Result

where the elements off the diagonal are equal to 0. We let $\mathrm{Sp}_n(q)$ be the symplectic group with respect to the symplectic form having matrix

$$\mathcal{J} := \begin{pmatrix} 0 & J \\ J & 0 \end{pmatrix}.$$

We construct the matrices u_+ and u_- explicitly. We let

$$A_1 := \begin{pmatrix} 1 & 1 & \cdots & \cdots & 1 \\ & 1 & 1 & \cdots & 1 \\ & & \ddots & \ddots & \vdots \\ & & & 1 & 1 \\ & & & & 1 \end{pmatrix}, \quad A_2 := \begin{pmatrix} 1 & 1 & & & \\ & 1 & 1 & & \\ & & \ddots & \ddots & \\ & & & 1 & 1 \\ & & & & 1 \end{pmatrix} \in \mathrm{GL}_m(q),$$

where as usual unmarked entries are tacitly assumed to be 0. Moreover, we let

$$j_1 := \begin{pmatrix} 1 \\ \vdots \\ 1 \end{pmatrix}, \quad j_2 := \begin{pmatrix} 1 \\ \vdots \\ 1 \\ 0 \end{pmatrix} \in \mathbb{F}_q^m.$$

Now we let

$$B_+ := \begin{pmatrix} j_1 & 0 & \cdots & 0 \end{pmatrix}, \quad B_- := \begin{pmatrix} j_1 & j_2 & 0 & \cdots & 0 \end{pmatrix} \in \mathrm{Mat}(m \times m, q).$$

Finally, we define

$$u_+ := \begin{pmatrix} A_1 & B_+ \\ 0 & A_2 \end{pmatrix}, \quad u_+ := \begin{pmatrix} A_1 & B_- \\ 0 & A_2 \end{pmatrix} \in \mathrm{SL}_n(q).$$

For instance, when $n = 8$, we have

$$u_+ = \begin{pmatrix} 1&1&1&1&1&0&0&0 \\ 0&1&1&1&1&0&0&0 \\ 0&0&1&1&1&0&0&0 \\ 0&0&0&1&1&0&0&0 \\ 0&0&0&0&1&1&0&0 \\ 0&0&0&0&0&1&1&0 \\ 0&0&0&0&0&0&1&1 \\ 0&0&0&0&0&0&0&1 \end{pmatrix}, \quad u_- = \begin{pmatrix} 1&1&1&1&1&1&0&0 \\ 0&1&1&1&1&1&0&0 \\ 0&0&1&1&1&1&0&0 \\ 0&0&0&1&1&0&0&0 \\ 0&0&0&0&1&1&0&0 \\ 0&0&0&0&0&1&1&0 \\ 0&0&0&0&0&0&1&1 \\ 0&0&0&0&0&0&0&1 \end{pmatrix}.$$

Observe that u_+ and u_- consists of a unique Jordan block and hence u_+ and u_- are regular unipotent elements.

Using the matrix defining the symplectic form for $\mathrm{Sp}_n(q)$, it is easy to verify that $u_+, u_- \in \mathrm{Sp}_n(q)$.

Now let E_{11} be the $m \times m$-matrix having 1 in position $(1, 1)$ and 0 anywhere else. Let

$$Q_+ := \begin{pmatrix} 0 & J \\ 0 & E_{11} \end{pmatrix} \in \mathrm{Mat}(n \times n, q).$$

Since $Q_+ + Q_+^T = \mathcal{J}$ is the matrix defining the symplectic form for $\mathrm{Sp}_n(q)$, we see that Q_+ defines a non-degenerate quadratic form polarizing to $\mathcal{J}$. We still denote with Q_+ this quadratic form. Observe that

$$Q_+(x_1, \ldots, x_{2m}) = x_1 x_{2m} + x_2 x_{2m-1} + \cdots + x_m x_{m+1} + x_{m+1}^2.$$

Since the $\mathbb{F}_q$-subspace spanned by the standard vectors $e_1, \ldots, e_m$ is totally singular for Q_+, we see that Q_+ has Witt defect zero. Let $\mathrm{SO}_n^+(q)$ be the special orthogonal group with respect to Q_+. Thus $\mathrm{SO}_n^+(q) \leq \mathrm{Sp}_n(q)$. A matrix computation shows that u_+ preserves Q_+ and hence $u_+ \in Q_+$.

Now let $a \in \mathbb{F}_q$ be such that the degree 2 polynomial $T^2 + T + a \in \mathbb{F}_q[T]$ is irreducible. Now let E_{mm} be the $m \times m$-matrix having 1 in position (m, m) and 0 anywhere else. Let

$$Q_- := \begin{pmatrix} E_{mm} & J \\ 0 & aE_{11} \end{pmatrix} \in \mathrm{Mat}(n \times n, q).$$

Since $Q_- + Q_-^T = \mathcal{J}$ is the matrix defining the symplectic form for $\mathrm{Sp}_n(q)$, we see that Q_- defines a non-degenerate quadratic form polarizing to $\mathcal{J}$. We still denote with Q_- this quadratic form. Observe that

$$Q_-(x_1, \ldots, x_{2m}) = x_1 x_{2m} + x_2 x_{2m-1} + \cdots + x_{m-1} x_{m+2} + x_m x_{m+1} + x_m^2 + a x_{m+1}^2.$$

We claim that Q_- has Witt defect 1. Let $e_1, \ldots, e_{2m}$ be the standard basis for $\mathbb{F}_q^{2m}$. Then $W = \langle e_1, \ldots, e_{m-1}\rangle$ is a totally singular subspace of $\mathbb{F}_q^{2m}$ with respect to Q_-. Set $W' = \langle W, e_m, e_{m+1}\rangle$. Clearly, $W^T \leq W'$. The quadratic form Q_- induces on W'/W the quadratic form $x_m^2 + x_m x_{m+1} + a x_{m+1}^2$. Since the polynomial $T^2 + T + a \in \mathbb{F}_q[T]$ is irreducible, the quadratic form $x_m^2 + x_m x_{m+1} + a x_{m+1}^2$ on W'/W is anisotropic and hence W is a maximal totally isotropic subspace of $\mathbb{F}_q^{2m}$. Since $\dim_{\mathbb{F}_q}(W) = m - 1$ and $\dim_{\mathbb{F}_q}(\mathbb{F}_q^{2m}) = 2m$, we deduce that Q_- has Witt defect 1.

Let $\mathrm{SO}_n^-(q)$ be the special orthogonal group with respect to Q_-. Thus $\mathrm{SO}_n^-(q) \leq \mathrm{Sp}_n(q)$. A matrix computation shows that u_- preserves Q_- and hence $u_- \in \mathrm{SO}_n^-(q)$. In particular, part (2) holds true.

Part (4) follows from [30, Proposition 7.1.9] or [31, Proposition 5.1.7].

6.3 An Auxiliary Result

Since $u_+ \in SO_n^+(q)$ and $u_- \in SO_n^-(q)$, part (3) implies (1). Hence, it remains to prove part (3).

From [30, Proposition 7.1.9] or [31, Proposition 5.1.7], we deduce that no $Sp_n(q)$-conjugate of u_+ lies in $SO_n^-(q)$ and, similarly, no $Sp_n(q)$-conjugate of u_- lies in $SO_n^+(q)$. Let $\phi : \mathbb{F}_q \to \mathbb{F}_q$ be defined by $x \mapsto x^2$, $\forall x \in \mathbb{F}_q$. Now ϕ induces a field automorphism on $Sp_n(q)$ which, by abuse of notation, we still denote by ϕ. When $n \geq 6$, as $\mathrm{Aut}(Sp_n(q)) = Sp_n(q) \rtimes \langle \phi \rangle$ and as ϕ centralizes u_+ and u_-, part (3) follows. The same argument, when $n = 4$, shows that no $Sp_4(q) \rtimes \langle \phi \rangle$-conjugate of u_+ lies in $SO_4^-(q)$ and no $Sp_4(q) \rtimes \langle \phi \rangle$-conjugate of u_- lies in $SO_4^+(q)$. For the rest of the argument we suppose by contradiction that $u_+^g \in SO_4^-(q)$, for some $g \in \mathrm{Aut}(Sp_4(q)) \setminus Sp_4(q) \rtimes \langle \phi \rangle$.

From [9, Table 8.14], the group $SO_4^+(q)^g \in \mathcal{C}_2$ and hence, $SO_4(q)^g$ preserves a direct sum decomposition $\mathbb{F}_q^4 = W_1 \perp W_2$, with $\dim_{\mathbb{F}_q}(W_1) = \dim_{\mathbb{F}_q}(W_2) = 2$ and W_1, W_2 non-degenerate.

From [71, Table 1, p. 10], the group $Sp_4(q)$ admits a maximal factorization with $SO_4^-(q)$ and with a maximal subgroup in the Aschbacher class $\mathcal{C}_2$. Therefore, as $Sp_2(q) \mathrm{wr} 2 \cong SO_4^+(q)^g \in \mathcal{C}_2$, we deduce

$$Sp_4(q) = SO_4^-(q) SO_4^+(q)^g$$

and we may suppose $W_1 = \langle e_1, e_4 \rangle$ and $W_2 = \langle e_2, e_3 \rangle$. Since

$$|Sp_4(q)| = q^4(q^4-1)(q^2-1), \; |SO_4^+(q)| = 2q^2(q^2-1)^2 \text{ and}$$
$$|SO_4^-(q)| = 2q^2(q^4-1),$$

we deduce

$$|SO_4^-(q) \cap SO_4^+(q)^g| = \frac{2q^2(q^2-1)^2 \cdot 2q^2(q^4-1)}{q^4(q^4-1)(q^2-1)} = 4(q^2-1).$$

This implies that a Sylow 2-subgroup of $SO_4^-(q) \cap SO_4^+(q)^g$ has order 4.

Finally, it is not hard to verify that the two matrices

$$\begin{pmatrix} 0 & 0 & 0 & 1 \\ 0 & 1 & 0 & 0 \\ 0 & 0 & 1 & 0 \\ 1 & 0 & 0 & 0 \end{pmatrix}, \begin{pmatrix} 1 & 0 & 0 & 0 \\ 0 & 1 & 0 & 0 \\ 0 & 1 & 1 & 0 \\ 0 & 0 & 0 & 1 \end{pmatrix}$$

preserve the quadratic form Q_- and preserve the direct sum decomposition $\mathbb{F}_q^4 = \langle e_1, e_4 \rangle \perp \langle e_2, e_3 \rangle$. Therefore, these two matrices belong to $SO_4^-(q) \cap SO_4^+(q)^g$. This shows that a Sylow 2-subgroup of $SO_4^-(q) \cap SO_4^+(q)^g$ is elementary abelian of order 4.

Since u_+ has order 4, we contradict the fact that $u_+^g \in SO_4^-(q) \cap SO_4^+(q)^g$. □

Chapter 7
Orthogonal Groups with Witt Defect 1

In this chapter, we deal with even dimensional orthogonal groups $P\Omega_n^-(q)$ of Witt defect 1, for $n \geq 8$.

Note that the cases $n \in \{2, 4, 6\}$ are not considered here because $\Omega_2^-(q) \cong C_{\frac{q+1}{\gcd(2,q-1)}}$ is abelian, while $P\Omega_4^-(q) \cong \mathrm{PSL}_2(q^2)$ and $P\Omega_6^-(q) \cong \mathrm{PSU}_4(q)$ are treated in the previous sections.

We begin with some preliminary lemmas.

Lemma 7.1 *Let s be a Singer cycle for* $\mathrm{GL}_\ell(q)$, *with $\ell \geq 1$ and q odd, and let*

$$x_\ell := \begin{pmatrix} s & 0 \\ 0 & (s^{-1})^T \end{pmatrix} \in \mathrm{SO}_{2\ell}^+(q).$$

Then $\mathbf{o}(x_\ell) = q^\ell - 1$ *and* $x_\ell \notin \Omega_{2\ell}^+(q)$.

Proof We describe $\mathrm{SO}_{2\ell}^+(q)$ with respect to the orthogonal form having matrix

$$J := \begin{pmatrix} 0_\ell & I_\ell \\ I_\ell & 0_\ell \end{pmatrix},$$

with respect to the basis $(e_1, \ldots, e_\ell, e_{\ell+1}, \ldots, e_{2\ell})$ of $V = \mathbb{F}_q^\ell$. Let $W := \langle e_1, \ldots, e_\ell \rangle$ and $U := \langle e_{\ell+1}, \ldots, e_{2\ell} \rangle$. Note that x_ℓ fixes both the subspaces W and U. By Lemma 2.7.2 in [66], we have that $x_\ell \in \Omega_{2\ell}^+(q)$ if and only if $\det_W(x_\ell)$ is a square in $\mathbb{F}_q$. Now, by (2.3), we have

$$\det_W(x_\ell) = \det_{\mathbb{F}_q}(s) = a^{\frac{q^\ell-1}{q-1}},$$

where $\langle a \rangle = \mathbb{F}_{q^\ell}^*$. Note that $b := a^{\frac{q^\ell-1}{q-1}}$ generates $\mathbb{F}_q^*$. Now recall that, since q is odd, the subgroup $(\mathbb{F}_q^*)^2$ has index 2 in $\mathbb{F}_q^*$. Thus $b \notin (\mathbb{F}_q^*)^2$ and hence $x_\ell \notin \Omega_{2\ell}^+(q)$. □

Lemma 7.2 *Let $n \geq 4$ be an even positive integer and let $1 \leq \ell \leq n/2 - 1$. Then $\Omega_n^-(q)$ contains an element having order*

$$\frac{(q^\ell + 1)(q^{\frac{n}{2}-\ell} - 1)}{\gcd\left(q^\ell + 1, q^{\frac{n}{2}-\ell} - 1\right)}$$

and action type $2\ell \oplus \left(\frac{n}{2} - \ell\right) \oplus \left(\frac{n}{2} - \ell\right)$.

Proof Let s be a Singer cycle of $\mathrm{SO}_{2\ell}^-(q)$; in particular, $\mathbf{o}(s) = q^\ell + 1$. Let $x_{\frac{n}{2}-\ell} \in \mathrm{SO}_{n-2\ell}^+(q) \setminus \Omega_{n-2\ell}^+(q)$ having order $\mathbf{o}(x_{\frac{n}{2}-\ell}) = q^{n/2-\ell} - 1$ and action type $\left(\frac{n}{2} - \ell\right) \oplus \left(\frac{n}{2} - \ell\right)$ as described in Lemma 7.1. Set $x := s \oplus x_{\frac{n}{2}-\ell} \in \mathrm{SO}_n^-(q)$. Then x has the required order and action type and we only need to show that $x \in \Omega_n^-(q)$. When q is even, as $\mathbf{o}(x)$ is odd and $|\mathrm{SO}_n^-(q) : \Omega_n^-(q)|$ is a power of 2, we deduce $x \in \Omega_n^-(q)$. Suppose q odd. Observe that $s \in \mathrm{SO}_{2\ell}^-(q) \setminus \Omega_{2\ell}^-(q)$ because the order of a Singer cycle in $\Omega_{2\ell}^-(q)$ for q odd is $(q^\ell + 1)/2$. Hence the spinor norm of s and $x_{\frac{n}{2}-\ell}$ is non-trivial and hence the spinor norm of $x = s \oplus x_{\frac{n}{2}-\ell}$ is trivial, that is, $x \in \Omega_n^-(q)$. □

Lemma 7.3 ([74, Theorem 1.1]) *Let n be an even integer with $n \geq 8$ and let M be a maximal subgroup of $\Omega_n^-(q)$ containing a Singer cycle. Then one of the following holds*

(1) $M \cong \Omega_{n/s}^-(q^s).s$ *is in class $\mathcal{C}_3$ and s is a prime divisor of $n/2$ with $n/s \geq 4$,*
(2) $n/2$ *is odd and M is of type $\mathrm{GU}_{n/2}(q) \in \mathcal{C}_3$.*

The detailed structure of M in part (2) can be found in [66, Proposition 4.3.18].

Proof From [74, Theorem 1.1], it follows that M is in the Aschbacher class $\mathcal{C}_3$. Moreover, from [66, Table 4.3. A], we deduce that either part (1) or (2) holds, or $n/2$ is odd, q is odd and $M \cong \Omega_{n/2}(q^2).2$. We show that $\Omega_{n/2}(q^2).2$ in fact does not contain a Singer cycle of $\Omega_n^-(q)$. Assume the contrary. From [51, Theorem 2.16] applied with $T = \Omega_{n/2}(q^2)$, the maximum element order of $\Omega_{n/2}(q^2).2$ is at most $q^{n/2+1}/(q^2 - 1)$. Since the order of a Singer cycle of $\Omega_n^-(q)$ is $(q^{n/2} + 1)/2$. We deduce

$$q^{n/2+1}/(q^2 - 1) \geq (q^{n/2} + 1)/2.$$

However, this is never satisfied. □

7.1 Small Even Dimensional Orthogonal Groups of Witt Defect 1

We start our analysis with small even dimensional orthogonal groups of Witt defect 1. For the subgroup structure of $P\Omega_8^-(q)$, $P\Omega_{10}^-(q)$ and $P\Omega_{12}^-(q)$, we use [9].

Lemma 7.4 *The weak normal covering number of $P\Omega_8^-(q)$ is at least 3.*

Proof As usual, we work with $\Omega_8^-(q)$. Consider a weak normal covering $\tilde{\mu}$ of $\Omega_8^-(q)$ of minimum cardinality and maximal components. Let $\tilde{H}$ be a component of $\tilde{\mu}$ containing a Singer cycle. From Lemma 7.3, we get

$$\tilde{H} \cong \Omega_4^-(q^2).2.$$

Recall from [66, Proposition 2.9.1] that $\Omega_4^-(q^2) \cong PSL_2(q^4)$.

Let y be an element of $\Omega_8^-(q)$ having order $(q^3-1)(q+1)/\gcd(q^3-1,q+1) = (q^3-1)(q+1)/\gcd(2,q-1)$ and action type $2 \oplus 3 \oplus 3$. The existence of such an element is guaranteed by Lemma 7.2.

Moreover, we have that $V = U_1 \oplus U_2 \oplus U_3$, with U_1, U_2, U_3 irreducible $\mathbb{F}_q\langle y\rangle$-modules, $\dim_{\mathbb{F}_q} U_1 = 2$, $\dim_{\mathbb{F}_q} U_2 = \dim_{\mathbb{F}_q} U_3 = 3$, the orthogonal form of $\Omega_8^-(q)$ restricted to U_1 is of Witt defect 1 and restricted to $U_2 \oplus U_3$ is of Witt defect 0 where U_2 and U_3 are totally isotropic. A case-by-case analysis on the maximal subgroups of $\Omega_8^-(q)$ in [9, Tables 8.52, 8.53] reveals that there are two $\Omega_8^-(q)$-conjugacy classes of maximal subgroups M containing a conjugate of y. Namely,

(1) $M \cong E_q^{3+6} : \left(\frac{1}{\gcd(2,q-1)}GL_3(q) \times \Omega_2^-(q)\right) \cdot \gcd(2,q-1) \in \mathcal{C}_1$,
(2) $M \cong (\Omega_2^-(q) \times \Omega_6^+(q)).2^{\gcd(2,q-1)} \in \mathcal{C}_1$.

In particular, since $\tilde{H}$ is not in this list we deduce $\gamma_w(\Omega_8^-(q)) \geq 2$. Let $\tilde{K}$ be a component of $\tilde{\mu}$ containing a conjugate of y. Thus, $\tilde{K}$ must be in one of the above two possibilities.

Let $z \in \Omega_8^-(q)$ having order $\frac{(q^3+1)(q-1)}{\gcd(2,q-1)}$ and action type $6 \oplus 1 \oplus 1$. The existence of such an element is guaranteed by Lemma 7.2.

The group $\tilde{H}$ does not contain elements having this order because $q^3 + 1$ does not divide $|\tilde{H}|$. This can be easily verified when $q \neq 2$ by showing that a primitive prime divisor of $q^6 - 1$ does not divide $|\tilde{H}| = 2(q^8 - 1)$ and it can be checked directly when $q = 2$. Moreover, none of the two possibilities above for $\tilde{K}$ contains elements of order $q^3 + 1$: this can be easily verified when $q \neq 2$ by showing that a primitive prime divisor of $q^6 - 1$ does not divide $|\tilde{K}|$ and it can be checked directly when $q = 2$. Therefore $\gamma_w(P\Omega_8^-(q)) \geq 3$. $\square$

Lemma 7.5 *The weak normal covering number of $P\Omega_{10}^-(q)$ is at least 3.*

Proof We have verified the veracity of this lemma with a computer when $q \in \{2,3\}$ and hence in the rest of the proof we assume $q \geq 4$. Let $\tilde{\mu}$ be a weak normal covering of $\Omega_{10}^-(q)$ of minimum cardinality and maximal components and let $\tilde{H}$ be a component containing a Singer cycle. From Lemma 7.3 and by [9], we obtain

$$\tilde{H} \cong \left(\frac{q+1}{\gcd(2, q+1)} \circ \mathrm{SU}_5(q)\right).\gcd(q+1, 5).$$

Let z be an element of $\Omega_{10}^-(q)$ such that

- $o(z) = (q^4+1)(q-1)/\gcd(q^4+1, q-1) = (q^4+1)(q-1)/\gcd(2, q-1)$,
- the action of z on V is of type $1 \oplus 1 \oplus 8$,
- $V = U_1 \oplus U_2 \oplus U_3$, with U_1, U_2, U_3 irreducible $\mathbb{F}_q\langle z\rangle$-modules, $\dim_{\mathbb{F}_q} U_1 = \dim_{\mathbb{F}_q} U_2 = 1$, $\dim_{\mathbb{F}_q} U_3 = 8$, the orthogonal form of $\Omega_{10}^-(q)$ restricted to $U_1 \oplus U_2$ is non-degenerate of Witt defect 0 (with U_1 and U_2 totally singular) and restricted to U_3 is non-degenerate of Witt defect 1.

The existence of such an element is guaranteed by Lemma 7.2.

We prove that $\tilde{H}$ contains no element having order divisible by q^4+1. Actually, we prove that the order of $\tilde{H}$ is not divisible by a primitive prime divisor $r \in P_8(q)$. Clearly r does not divide the size of $\frac{q+1}{\gcd(2,q+1)} \circ \mathrm{SU}_5(q)$. Moreover, by (2.4), we have that $r \geq 9$ so that $r \nmid \gcd(q+1, 5)$. Thus $r \nmid |\tilde{H}|$.

Therefore, no $\mathrm{Aut}(\Omega_{10}^-(q))$-conjugate of z lies in $\tilde{H}$.

Assume that $\gamma_w(\Omega_{10}^-(q)) = 2$ so that $\tilde{\mu} = \{\tilde{H}, \tilde{K}\}$, where $\tilde{K}$ is a maximal component containing an $\mathrm{Aut}(\Omega_{10}^-(q))$-conjugate of z. Our task is reaching a contradiction.

Using the fact that $q \geq 4$, a case-by-case analysis on the maximal subgroups of $\Omega_{10}^-(q)$ in [9, Tables 8.68, 8.69] reveals that the maximal subgroups M having order divisible by q^4+1 are, up to $\Omega_{10}^-(q)$-conjugacy, as follows

1. $M \cong E_q^8 : \left(\frac{q-1}{\gcd(2,q-1)} \times \Omega_8^-(q)\right).\gcd(2, q-1) \in \mathcal{C}_1$,
2. q is even and $M \cong \mathrm{Sp}_8(q) \in \mathcal{C}_1$
3. q is odd and $M \cong \Omega_9(q).2 \in \mathcal{C}_1$,
4. $M \cong (\Omega_2^+(q) \times \Omega_8^-(q)).2^{\gcd(2,q-1)} \in \mathcal{C}_1$,
5. $q \equiv 1 \pmod{4}$ and $M \cong \Omega_5(q^2).2 \in \mathcal{C}_3$,
6. $q \equiv 3 \pmod{4}$ and $M \cong 2 \times \Omega_5(q^2).2 \in \mathcal{C}_3$.

In particular, $\tilde{K}$ is one of these groups.

Let x be an element of $\Omega_{10}^-(q)$ such that

- $o(x) = (q^3-1)(q^2+1)/\gcd(2, q-1)$ and x has action type $4 \oplus 3 \oplus 3$;
- $V = U_1 \perp (U_2 \oplus U_3)$, with U_1, U_2, U_3 irreducible $\mathbb{F}_q\langle x\rangle$-modules, $\dim_{\mathbb{F}_q} U_1 = 4$, $\dim_{\mathbb{F}_q} U_2 = \dim_{\mathbb{F}_q} U_3 = 3$, the orthogonal form of $\Omega_{10}^-(q)$ restricted to U_1 is of Witt defect 1 and restricted to $U_2 \oplus U_3$ is of Witt defect 0 where U_2 and U_3 are totally isotropic;
- x induces on U_2 and U_3 a matrix of order $q^3 - 1$ and on U_1 a Singer cycle of order q^2+1.

The existence of such an element is guaranteed by Lemma 7.2.

7.1 Small Even Dimensional Orthogonal Groups of Witt Defect 1

Now, x does not fix non-zero totally isotropic subspaces of dimension less or equal to 2 and hence x has no $\mathrm{Aut}(\Omega_{10}^-(q))$-conjugate in

- $M \cong E_q^8 : \left(\frac{q-1}{\gcd(2,q-1)} \times \Omega_8^-(q)\right) . \gcd(2, q-1) \in \mathcal{C}_1$, or in
- $M \cong \mathrm{Sp}_8(q) \in \mathcal{C}_1$ with q even.

Similarly, x does not fix a non-degenerate subspace of dimension less or equal to 2 and hence x has no $\mathrm{Aut}(\Omega_{10}^-(q))$-conjugate in

- $M \cong \Omega_9(q).2 \in \mathcal{C}_1$ with q odd, or in
- $M \cong (\Omega_2^+(q) \times \Omega_8^-(q)).2^{\gcd(2,q-1)} \in \mathcal{C}_1$.

In the remaining two possibilities for $\tilde{K}$ we have that

$$|\tilde{K}| \in \{2(q^4 - 1)(q^8 - 1), 4(q^4 - 1)(q^8 - 1)\}.$$

We show that $q^3 - 1$ does not divide $4(q^4 - 1)(q^8 - 1)$, so that $\tilde{K}$ does not contain x or any of its conjugates. Let $r \in P_3(q)$. Then, by (2.4), $r \geq 5$ so that $r \nmid 4(q^4 - 1)(q^8 - 1)$. By the above considerations on $r \in P_3(q)$, we also have that $r \nmid (q^2 - 1)(q^4 - 1)(q^5 + 1)$. Moreover, by Lemma 2.2, we also have $r \nmid q^3 + 1$ and hence x has no $\mathrm{Aut}(\Omega_{10}^-(q))$-conjugate in $\tilde{H}$.

Therefore $\gamma_w(\mathrm{P}\Omega_{10}^-(q)) \geq 3$. □

Lemma 7.6 *The weak normal covering number of* $\mathrm{P}\Omega_{12}^-(q)$ *is at least* 3.

Proof We have verified the veracity of this lemma with a computer when $q \in \{2, 3\}$ and hence in the rest of the proof we assume $q \geq 4$. As usual we may work with $\Omega_{12}^-(q)$. Let $\tilde{\mu}$ be a weak normal covering of $\Omega_{12}^-(q)$ of minimum cardinality and maximal components and let $\tilde{H}$ be a component containing a Singer cycle. From Lemma 7.3, we obtain

$$\tilde{H} \cong \Omega_4^-(q^3).3 \text{ or } \tilde{H} \cong \Omega_6^-(q^2).2.$$

Let $z \in \Omega_{12}^-(q)$ having order $\frac{(q^5+1)(q-1)}{\gcd(2,q-1)}$ and action type $10 \oplus 1 \oplus 1$. The existence of such an element is guaranteed by Lemma 7.2.

The group $\tilde{H}$ does not contain elements having this order because $q^5 + 1$ does not divide $|\tilde{H}|$: this can be easily verified using the structure of $\tilde{H}$ and considering a primitive prime divisor of $q^{10} - 1$. In particular, $\gamma_w(\Omega_{12}^-(q)) \geq 2$.

Let $\tilde{K}$ be a component of $\tilde{\mu}$ containing an $\mathrm{Aut}(\Omega_{12}^-(q))$-conjugate of z. Using the fact that $q \geq 4$, a case-by-case analysis on the maximal subgroups of $\Omega_{12}^-(q)$ in [9, Tables 8.84, 8.85] reveals that the maximal subgroups M having order divisible by $q^5 + 1$ are as follows

(1) $M \cong E_q^{10} : \left(\frac{q-1}{\gcd(2,q-1)} \times \Omega_{10}^-(q)\right) . \gcd(2, q-1) \in \mathcal{C}_1$,
(2) q is even and $M \cong \mathrm{Sp}_{10}(q) \in \mathcal{C}_1$
(3) q is odd and $M \cong \Omega_{11}(q).2 \in \mathcal{C}_1$,
(4) $M \cong (\Omega_2^+(q) \times \Omega_{10}^-(q)).2^{\gcd(2,q-1)} \in \mathcal{C}_1$.

In particular, $\tilde{K}$ is one of these groups and in all cases $\tilde{K}$ is in the Aschbacher class $\mathcal{C}_1$. Furthermore, in the first two cases $\tilde{K}$ fixes a non-zero totally isotropic subspace of dimension at most 2 and in the last two cases $\tilde{K}$ fixes a non-degenerate subspace of dimension at most 2.

Assume $\gamma_w(\Omega_{12}^-(q)) = 2$. Let y be an element of $\Omega_{12}^-(q)$ having order

$$\frac{(q^3-1)(q^3+1)}{\gcd(q^3-1, q^3+1)} = \frac{q^6-1}{\gcd(2, q-1)},$$

action type $6 \oplus 3 \oplus 3$ and decomposing $V = U_1 \perp (U_2 \oplus U_3)$, with U_1, U_2, U_3 irreducible $\mathbb{F}_q\langle y\rangle$-modules, $\dim_{\mathbb{F}_q} U_1 = 6$, $\dim_{\mathbb{F}_q} U_2 = \dim_{\mathbb{F}_q} U_3 = 3$, the orthogonal form of $\Omega_{12}^-(q)$ restricted to U_1 has Witt defect 1 and restricted to $U_2 \oplus U_3$ has Witt defect 0, where U_2 and U_3 are totally isotropic. The existence of such an element is guaranteed by Lemma 7.2.

Now, y does not fix non-zero subspaces of dimension ≤ 2 and hence y has no $\mathrm{Aut}(\Omega_{12}^-(q))$-conjugate in $\tilde{K}$. On the other hand, by the usual considerations on a primitive prime divisor r of $q^6 - 1$, we see that r divides the order of y but does not divide the order of $\Omega_6^-(q^2).2$. Note that in the critical case for Zsigmondy's theorem, that is, $q^6 - 1$ has no primitive prime divisors, we have that $o(y) = 63$ and $7 \nmid |\Omega_6^-(4).2|$. Since $\gamma_w(\Omega_{12}^-(q)) = 2$, we have

$$\tilde{H} \cong \Omega_4^-(q^3).3.$$

Let x be an element of $\Omega_{12}^-(q)$ having order

$$\frac{(q^5-1)(q+1)}{\gcd(q^5-1, q+1)} = \frac{(q^5-1)(q+1)}{\gcd(2, q-1)},$$

as in Lemma 7.2. In particular, x has action type $5 \oplus 5 \oplus 2$ and decomposes $V = U_1 \perp (U_2 \oplus U_3)$, with U_1, U_2, U_3 irreducible $\mathbb{F}_q\langle x\rangle$-modules.

Now, x does not fix non-zero totally isotropic subspaces of dimension at most 2 and hence x has no $\mathrm{Aut}(\Omega_{12}^-(q))$-conjugate in

- $M \cong E_q^{10} : \left(\frac{q-1}{\gcd(2,q-1)} \times \Omega_{10}^-(q)\right).\gcd(2, q-1) \in \mathcal{C}_1$, or in
- $M \cong \mathrm{Sp}_{10}(q) \in \mathcal{C}_1$ with q even.

Similarly, x does not fix a non-degenerate subspace of dimension 1 and hence x has no $\mathrm{Aut}(\Omega_{12}^-(q))$-conjugate in

- $M \cong \Omega_{11}(q).2 \in \mathcal{C}_1$ with q odd.

Finally, the only 2-dimensional non-degenerate subspace of V fixed by x is of "minus type" by construction and hence x has no $\mathrm{Aut}(\Omega_{12}^-(q))$-conjugate in

- $M \cong (\Omega_2^+(q) \times \Omega_{10}^-(q)).2^{\gcd(2,q-1)} \in \mathcal{C}_1$.

Therefore, regardless of the structure of $\tilde{K}$, we have shown that x has no $\mathrm{Aut}(\Omega_{12}^-(q))$-conjugate in $\tilde{K}$.

On the other hand, the group $\tilde{H} \cong \Omega_4^-(q^3).3$ does not contain elements having the order of x, because $q^5 - 1$ does not divide $|\tilde{H}|$ by the usual considerations on a primitive prime divisor of $q^5 - 1$. Therefore $\gamma_w(P\Omega_{12}^-(q)) \geq 3$. □

7.2 Large Even Dimensional Orthogonal Groups of Witt Defect 1

We start this section with a technical lemma; in what follows, we only need a few special cases of Lemma 7.7.

Lemma 7.7 *Let m be an integer with $m \geq 6$, let ℓ be odd with $1 \leq \ell \leq m$ and $\gcd(m, \ell) = 1$, and let $x \in \Omega_{2m}^-(q)$ with*

$$o(x) = \frac{(q^\ell - 1)(q^{m-\ell} + 1)}{\gcd(q^\ell - 1, q^{m-\ell} + 1)}.$$

If $x \in M$ with M a maximal subgroup of $\Omega_{2m}^-(q)$ in class $\mathcal{C}_3$, then m is odd, $\ell < m/2$ and $M \cong \Omega_m(q^2).2$.

Note that the existence of at least one $x \in \Omega_{2m}^-(q)$ of the above order is guaranteed by Lemma 7.2.

Proof From Lemma 2.2 (3), we have $\gcd(q^\ell - 1, q^{m-\ell} + 1) = \gcd(2, q - 1)$ because ℓ is odd. Therefore,

$$o(x) = \frac{(q^\ell - 1)(q^{m-\ell} + 1)}{\gcd(2, q - 1)}.$$

Suppose that $x \in M$, where M is a maximal subgroup of $\Omega_{2m}^-(q)$ in the class $\mathcal{C}_3$. From [66, Section 4.3], we have three cases to consider:

- m is odd and M is of type $\mathrm{GU}_m(q)$,
- $M \cong \Omega_{2m/s}^-(q^s).s$, where s is a prime divisor of m with $m/s \geq 2$,
- m is odd and $M \cong \Omega_m(q^2).2$.

We show that the first two cases cannot occur and, if the third case does occur, then $\ell < m/2$. We divide the proof into two cases, depending on whether $\ell > m/2$ or $\ell < m/2$. Observe that the case $\ell = m/2$ is excluded because $\gcd(m, \ell) = 1$ and $m \geq 6$.

Case $\ell > m/2$
As ℓ is odd and $\ell > m/2 \geq 3$, from Zsigmondy's theorem, $P_\ell(q) \neq \emptyset$ and hence x is a strong $ppd(2m, q; \ell)$-element. Let $r \in P_\ell(q)$. Then r divides $|M|$ and, by (2.4), $r \geq \ell + 1 \geq 5$.

Suppose first that m is odd and M is of type $\mathrm{GU}_m(q)$. As above, we have

$$M \cong \frac{q+1}{a} \cdot \mathrm{PSU}_m(q) \cdot \gcd(q+1, m),$$

where $a := \gcd(q+1, 4)$. Thus

- r divides $q^k + 1$, for some odd k with $1 \leq k \leq m$, or
- r divides $q^k - 1$ for some even k with $1 \leq k \leq m$.

In the first case, by Lemma 2.2 (3), r divides $\gcd(q^\ell - 1, q^k + 1) = \gcd(2, q-1)$, against $r \geq 5$. In the second case, by Lemma 2.3 (3), ℓ divides k. As ℓ is odd and k is even, we deduce that 2ℓ divides k. Thus $m < 2\ell \leq k \leq m$, which is a contradiction. Therefore, this case does not arise.

Suppose next that $M \cong \Omega^-_{2m/s}(q^s).s$, for some prime divisor s of m with $m/s \geq 2$. Thus

- r divides $q^m + 1$, or
- r divides $q^{2sk} - 1$ for some k with $1 \leq k \leq m/s - 1$, or
- $r = s$.

If r divides $q^m + 1$, then using the fact that ℓ is odd, by Lemma 2.2 (3), we deduce that r divides $\gcd(q^\ell - 1, q^k + 1) = \gcd(2, q-1)$, which is impossible.

If r divides $q^{2sk} - 1$, then by Lemma 2.3 (3), we have $\ell \mid 2sk$. As ℓ is odd, we get $\ell \mid sk$. Since $sk < m$ and $\ell > m/2$, we get $sk = \ell$. Thus $s \mid \gcd(m, \ell) = 1$, which is a contradiction.

We now exclude also the possibility $r = s$. Since $\ell > 1$ and $r \in P_\ell(q)$, we deduce that r is odd. Again, from $r \in P_\ell(q)$, we deduce $\ell \mid r - 1$ and hence $r - 1 = \alpha \ell$, for some positive integer α. Since r and ℓ are odd, α is even. Therefore,

$$m - 1 \geq s - 1 = r - 1 = \alpha \ell \geq 2\ell > m,$$

which is a contradiction.

Suppose finally that $M \cong \Omega_m(q^2).2$ with m odd. Since $r \neq 2$ and since r is relatively prime to q, we have that

$$r \text{ divides } q^{4k} - 1 \text{ for some } k \text{ with } 1 \leq k \leq (m-1)/2.$$

Then, by Lemma 2.3 (3), we have $\ell \mid 4k$. As ℓ is odd, we get $\ell \mid k$ and hence $m/2 < \ell \leq k \leq (m-1)/2$, which is clearly a contradiction.

Case $\ell < m/2$

As $m \geq 6$ and $1 \leq \ell < m/2$, we have $2m - 2\ell \geq m + 1 \geq 7$ and thus, from Zsigmondy's theorem, $P_{2m-2\ell}(q) \neq \emptyset$ and hence x is a strong $ppd(2m, q; 2m - 2\ell)$-element. Let $r \in P_{2m-2\ell}(q)$. Then r divides $|M|$.

7.2 Large Even Dimensional Orthogonal Groups of Witt Defect 1

Suppose first that m is odd and M is of type $\mathrm{GU}_m(q)$. From [66, Proposition 4.3.18], we have

$$M \cong \frac{q+1}{a} \cdot \mathrm{PSU}_m(q) . \gcd(q+1, m),$$

where $a := \gcd(q+1, 4)$. Thus

- r divides $q^k + 1$, for some odd k with $1 \le k \le m$, or
- r divides $q^k - 1$ for some even k with $1 \le k \le m$.

In the first case, as r is a primitive prime divisor of $q^{2m-2\ell} - 1$, by Lemma 2.3 (3), we get $2m - 2\ell \mid 2k$ so that the even number $m - \ell$ divides the odd number k, which is absurd.

In the second case, using again the fact that r is a primitive prime divisor of $q^{2m-2\ell} - 1$, we get $k \ge 2m - 2\ell > m$, against $k \le m$.

Therefore, this case does not arise.

Suppose next that $M \cong \Omega^-_{2m/s}(q^s).s$, for some prime divisor s of m with $m/s \ge 2$. Thus

- r divides $q^m + 1$, or
- r divides $q^{2sk} - 1$ for some k with $1 \le k \le m/s - 1$, or
- $r = s$.

When r divides $q^m + 1$, note that by Lemma 2.3 (3), we have $2m - 2\ell \mid 2m$ and thus $m - \ell \mid m$. Since $m - \ell > m/2$, this implies $m - \ell = m$, which is clearly impossible.

When r divides $q^{2sk} - 1$, observe that by Lemma 2.3 (3), we have $2m - 2\ell \mid 2sk$ and thus $m - \ell \mid sk \le m$. Since $m - \ell > m/2$, this implies $m - \ell = sk$. As s divides m, we deduce that s divides ℓ. However, as $\gcd(\ell, m) = 1$, this yields $s = 1$, contradicting the fact that s is a prime number.

When $r = s$, from (2.4), we get $s \ge 2(m - \ell) + 1 > m + 1$ against the fact that $s \le m$.

Finally, when m is odd and $M \cong \Omega_m(q^2).2$, we reach the conclusion of the lemma. □

Proposition 7.8 *For every $n \ge 14$, the weak normal covering number of $\mathrm{P}\Omega^-_n(q)$ is at least 3.*

Proof As usual we may argue with $\Omega^-_n(q)$. Let $\tilde{H}$ be a maximal component of a weak normal covering of $\Omega^-_n(q)$ containing a Singer cycle. From Lemma 7.3, we have that $\tilde{H}$ is in class $\mathcal{C}_3$ and

- $\tilde{H} \cong \Omega^-_{n/s}(q^s).s$ for some prime divisor s of $n/2$ with $n/s \ge 4$, or
- $\tilde{H}$ is of type $\mathrm{GU}_{\frac{n}{2}}(q)$ and $n/2$ is odd.

Let $x \in \Omega_n^-(q)$ with

$$\mathbf{o}(x) = \frac{(q^{\frac{n}{2}-1}+1)(q-1)}{\gcd(q^{\frac{n}{2}-1}+1, q-1)} = \frac{(q^{\frac{n}{2}-1}+1)(q-1)}{\gcd(2, q-1)}$$

and action type $(n-2) \oplus 1 \oplus 1$ on the natural module V. The existence of such an element is guaranteed by Lemma 7.2.

For later, we need to observe that, by Sect. 2.5, the quadratic form of $\Omega_n^-(q)$ restricted to the $(n-2)$-dimensional x-invariant subspace of V has Witt defect 1. As $\tilde{H}$ is in class $\mathcal{C}_3$, Lemma 7.7 applied with $\ell := 1$ yields that $\tilde{H}$ contains no elements having order $\mathbf{o}(x)$.

Therefore, there exists a second component, $\tilde{K}$ say, containing an $\mathrm{Aut}(\Omega_n^-(q))$-conjugate of x. This shows $\gamma_w(\Omega_n^-(q)) \geq 2$. For the rest of the proof, we suppose that $\gamma_w(\Omega_n^-(q)) = 2$ and that $\tilde{H}, \tilde{K}$ are the maximal components of a weak normal 2-covering.

As $n \geq 14$, $P_{n-2}(q) \neq \emptyset$ and hence

$$\tilde{K} \text{ is a } ppd(n, q; n-2)\text{-group.}$$

Let ℓ be a prime number with $n/4 < \ell < n/2$. Observe that the existence of ℓ is guaranteed by Bertrand's postulate and that obviously ℓ is odd. Let $y \in \Omega_n^-(q)$ with

$$\mathbf{o}(y) = \frac{(q^{\frac{n}{2}-\ell}+1)(q^\ell-1)}{\gcd(q^{\frac{n}{2}-1}, q^\ell-1)} = \frac{(q^{\frac{n}{2}-\ell}+1)(q^\ell-1)}{\gcd(2, q-1)}$$

and action type $\ell \oplus \ell \oplus (n-2\ell)$. The existence of such an element is guaranteed by Lemma 7.2.

As $\tilde{H}$ is in class $\mathcal{C}_3$ and $\ell > n/4$, it follows from Lemma 7.7 that y does not have a conjugate in $\tilde{H}$. Hence y has an $\mathrm{Aut}(\Omega_n^-(q))$-conjugate in $\tilde{K}$. From Zsigmondy's theorem, we have that $P_\ell(q) \neq \emptyset$, y is a strong $ppd(n, q; \ell)$-element and

$$\tilde{K} \text{ is a } ppd(n, q; \ell)\text{-group.}$$

As $n - \ell > 3$, by Theorem 2.6, Lemma 2.7 (3) and Table 3.5. F in [66], $\tilde{K}$ belongs to one of the classes $\mathcal{C}_i$ for some $i \in \{1, 2, 3\}$ or to $\mathcal{S}$ and it is described in the Example 2.6 a) of [53].

The Subgroup $\tilde{K}$ Lies in Class $\mathcal{C}_1$ Here $\tilde{K}$ is the stabilizer of a totally singular or of a non-degenerate subspace of V.

Recall $V = U_1 \oplus U_2 \oplus U_3$, with U_1, U_2, U_3 irreducible $\mathbb{F}_q\langle x \rangle$-submodules of V, $\dim_{\mathbb{F}_q} U_1 = n-2$, $\dim_{\mathbb{F}_q} U_2 = \dim_{\mathbb{F}_q} U_3 = 1$ and with the restriction of the quadratic form on U_1 having Witt defect 1.

7.2 Large Even Dimensional Orthogonal Groups of Witt Defect 1

Now, we use the fact that $\tilde{K}$ contains an $\mathrm{Aut}(\Omega_n^-(q))$-conjugate of x. From the paragraph above, we deduce that one of the following holds

- $\tilde{K}$ is a parabolic subgroup $P_1 \in \mathcal{C}_1$,
- q is even and $\tilde{K} \cong \mathrm{Sp}_{n-2}(q) \in \mathcal{C}_1$,
- q is odd and $\tilde{K}$ is of type $\mathrm{O}_1(q) \perp \mathrm{O}_{n-1}(q) \in \mathcal{C}_1$,
- $\tilde{K}$ is of type $\mathrm{O}_2^+(q) \perp \mathrm{O}_{n-2}^-(q) \in \mathcal{C}_1$.

However, since $\ell, 2\ell, n - 2\ell, n - \ell \notin \{1, 2, n - 1, n - 2\}$, y stabilizes no subspaces of V of dimension 1 or 2. Therefore, this case does not arise.

The Subgroup $\tilde{K}$ Lies in Class $\mathcal{C}_2$ These are the groups in the Example 2.3 of [53]. Hence $\tilde{K} \leq \mathrm{GL}_1(q) \mathrm{wr} S_n$ is a $ppd(n, q; n - 2)$-group. Moreover, from [53, p. 170], $n - 1$ is a primitive prime divisor of $q^{n-2} - 1$, that is, $n - 1$ is a prime number with $n - 1 \geq 13$. Furthermore, by Proposition 4.2.15 in [66], we have that $q \equiv 3$ (mod 4) is a prime, $n \equiv 2$ (mod 4) and $\tilde{K} \leq E_2^n . S_n$. Let $n = 2 + 4k$ and note that, since $n \geq 14$, we have $k \geq 3$. By the fact that $\tilde{K}$ contains x, it follows that $\tilde{K}$ contains an element of order $q^{\frac{n}{2}-1} + 1$. Obviously $n - 1 \mid q^{\frac{n}{2}-1} + 1 = q^{2k} + 1$. Now note that $q^{2k} + 1 \equiv 3^{2k} + 1$ (mod 4) $\equiv 2$ (mod 4) so that 2 is the maximum power of 2 dividing $q^{2k} + 1$. Moreover, $q^{2k} + 1 > 2(n - 1) = 8k + 2$ because we have

$$q^{2k} + 1 \geq 3^{2k} + 1 > 8k + 2,$$

for every $k \geq 3$.

Thus there exists an odd prime r with $r(n - 1) \mid q^{2k} + 1$, and hence such that $r(n - 1) \mid \mathbf{o}(x)$. Hence we also have an element of order $r(n - 1)$ in $E_2^n . S_n$. It follows that there is an element of such an order in S_n. Recall that $n - 1$ is prime, see [53, Example 2.3]. When $r \neq n - 1$, this implies $n \geq r + n - 1 \geq n + 2$, which is a contradiction; when $r = n - 1$, this implies $(n - 1)^2 \leq n$, which is again a contradiction.

The Subgroup $\tilde{K}$ Lies in Class $\mathcal{C}_3$ Here we use the fact that $\tilde{K}$ contains an $\mathrm{Aut}(\Omega_n^-(q))$-conjugate of y. We conclude that this case cannot arise by Lemma 7.7, because $\ell > n/4$.

The Subgroup $\tilde{K}$ Lies in Class $\mathcal{S}$ Since $\tilde{K}$ is described in Example 2.6 a) of [53], we have

$$A_m \leq \tilde{K} \leq S_m \times \mathbf{Z}(\Omega_n^-(q)),$$

with $m \in \{n + 1, n + 2\}$. In particular, since $|\mathbf{Z}(\Omega_n^-(q))|$ divides 2 by [66, Table 2.1.D], we deduce $x^2 \in S_m$. In particular, S_m contains an element having order

$$\mathbf{o}(x^2) = \frac{\mathbf{o}(x)}{2} = \frac{(q^{\frac{n}{2}-1} + 1)(q - 1)}{4},$$

when q is odd, and having order

$$\mathbf{o}(x^2) = \mathbf{o}(x) = (q^{\frac{n}{2}-1} + 1)(q - 1),$$

when q is even. In both cases, when $q \neq 3$, we have that S_m contains an element of order at least $q^{\frac{n}{2}-1} + 1$; when $q = 3$, we have that S_m contains an element of order $(q^{\frac{n}{2}-1} + 1)/2$. Set $c := 1$ when $q \neq 3$ and $c := 2$ when $q = 3$. Arguing as in the symplectic case, we deduce

$$\log\left(\frac{q^{\frac{n}{2}-1}+1}{c}\right) \leq \sqrt{m \log m} \left(1 + \frac{\log(\log(m)) - 0.975}{2 \log(m)}\right). \tag{7.1}$$

This inequality holds true only[1] when $n \leq 36$ and $q = 2$, or $q = 3$ and $n \leq 16$. We have checked these cases with a computer and, when $m \in \{n+1, n+2\}$, S_m contains no elements having order $2^{\frac{n}{2}-1} + 1$ when $14 \leq n \leq 36$ and having order $(3^{\frac{n}{2}-1} + 1)/2$ when $14 \leq n \leq 16$. Therefore, also this case does not arise.

Summing up, when $n \geq 8$, $P\Omega_n^-(q)$ has weak normal covering number at least 3. $\square$

---

[1] The argument here is again similar to the proof of Lemma 5.7. The function $q \mapsto (q^{n/2-1} + 1)/c$ is increasing in q and hence $\log((q^{n/2-1} + 1)/c) \geq \log(2^{n/2-2}) = (n/2 - 2) \log(2)$. Moreover, as in the proof of Lemma 5.7, we have $\sqrt{m \log m} \left(1 + \frac{\log(\log(m)) - a}{2 \log(m)}\right) \leq 2\sqrt{(n+2) \log(n+2)}$.

It is now easy to verify that $(n/2 - 2) \log(2) > 2\sqrt{(n+2) \log(n+2)}$ when $n \geq 190$. We have implemented in a computer the inequality in (7.1) for dealing with small values of n.

Chapter 8
Orthogonal Groups with Witt Defect 0

In this chapter, we deal with even dimensional orthogonal groups $P\Omega_n^+(q)$ of Witt defect 0, with $n \geq 8$. Note that the case $n = 6$ is considered in Chap. 4 because $P\Omega_6^+(q) \cong \mathrm{PSL}_4(q)$. The cases $n \in \{2, 4\}$ are not considered because the groups $P\Omega_4^+(q)$ and $P\Omega_2^+(q)$ are not simple.

When $n \neq 8$, we may work with $\Omega_n^+(q)$ because the automorphism group of $\Omega_n^+(q)$ projects onto the automorphism group of $P\Omega_n^+(q)$. The case $n = 8$ is rather special. Triality automorphisms of $P\Omega_8^+(q)$ cannot be lifted to automorphisms of $\Omega_8^+(q)$ (unless q is even, because in that case $\Omega_8^+(q) = P\Omega_8^+(q)$). Moreover, triality automorphisms do not act on V and hence, for instance, an element and its image under a triality automorphism might have a rather different action on V.

Before embarking in the proofs of our main result, we define a family of elements in $\Omega_n^+(q)$. Given an even integer m with $2 \leq m \leq n/2$, we consider the embedding of $\mathrm{SO}_m^-(q) \perp \mathrm{SO}_{n-m}^-(q)$ in $\mathrm{SO}_n^+(q)$. Let $s_m \in \mathrm{SO}_m^-(q)$ and $s_{n-m} \in \mathrm{SO}_{n-m}^-(q)$ be Singer cycles having order $q^{m/2} + 1$ and $q^{n/2-m/2} + 1$, respectively. By Proposition 2.8, we have that $x := s_m \oplus s_{n-m} \in \Omega_n^+(q)$ has action type $m \oplus (n - m)$ and

$$\mathbf{o}(x) = \frac{(q^{\frac{m}{2}} + 1)(q^{\frac{n}{2}-\frac{m}{2}} + 1)}{\gcd(q^{\frac{m}{2}} + 1, q^{\frac{n}{2}-\frac{m}{2}} + 1)}.$$

We will need the element x defined above with $m = 2$ in Lemma 8.1 and with $m = 4$ in Lemma 8.2. Note that, by Lemma 2.2, when $m = 2$, we have

$$\mathbf{o}(x) = \frac{(q+1)(q^{\frac{n}{2}-1}+1)}{\gcd(q+1, q^{\frac{n}{2}-1}+1)} = \begin{cases} q^{\frac{n}{2}-1} + 1 & \text{if } \frac{n}{2} \text{ is even,} \\ \frac{(q+1)(q^{\frac{n}{2}-1}+1)}{\gcd(2,q-1)} & \text{if } \frac{n}{2} \text{ is odd,} \end{cases}$$

while, when $m = 4$, we have

$$o(x) = \frac{(q^2+1)(q^{\frac{n}{2}-2}+1)}{\gcd(q^2+1, q^{\frac{n}{2}-2}+1)} = \frac{(q^2+1)(q^{\frac{n}{2}-2}+1)}{\gcd(2, q-1)}.$$

8.1 Orthogonal Groups of Witt Defect 0 Having Dimension At Least 10

Lemma 8.1 *Let M be a maximal subgroup of $\Omega_n^+(q)$ with $n \geq 10$ containing an element x of order $\frac{(q+1)(q^{\frac{n}{2}-1}+1)}{\gcd(q+1,q^{\frac{n}{2}-1}+1)}$ and action type $2 \oplus (n-2)$. Then, one of the following holds*

(1) $M \cong (\Omega_2^-(q) \perp \Omega_{n-2}^-(q)).2^{\gcd(2,q-1)} \in \mathcal{C}_1$,
(2) $n/2$ *is even and M is of type* $\mathrm{GU}_{n/2}(q).2 \in \mathcal{C}_3$.

For the detailed structure of M in part (2) see [66, Propositions 4.3.18, 4.3.20]. In particular, in (2), we stress here the fact that

$$|M| \text{ divides } |\mathrm{GU}_{n/2}(q).2|.$$

Proof We start by observing that [74, Theorem 1.1] describes the maximal subgroups of $\Omega_n^+(q)$ containing a low-Singer cycle s, that is, an element of $\Omega_n^+(q)$ of order $q^{n/2-1}+1$ and action $(n-2) \oplus 1 \oplus 1$ inducing a Singer cycle belonging to $O_{n-2}^-(q)$ on its invariant subspace of dimension $n-2$ and the identity matrix on its invariant subspace of dimension 2. Clearly both x in the statement of this lemma and s in [74, Theorem 1.1] are strong $ppd(n, q; n-2)$-elements. Now, the proof of [74, Theorem 1.1] just uses this fact. Hence we reach the same conclusion for the maximal subgroups containing s and those containing x: namely, they belong to $\mathcal{C}_1 \cup \mathcal{C}_3$. If $M \in \mathcal{C}_3$, then, using [74, Table II], we have that either part (2) holds or

$$nq/2 \text{ is odd and } M \text{ is of type } \Omega_{n/2}(q^2).2 \in \mathcal{C}_3. \tag{$\dagger$}$$

Furthermore, if $M \in \mathcal{C}_1$, then with a direct inspection on $\mathcal{C}_1$, we get (1). Therefore, to conclude the proof we need to exclude the case described in ($\dagger$): to do this we use the action type of x.

Suppose $nq/2$ is odd, M is of type $\Omega_{n/2}(q^2).2 \in \mathcal{C}_3$ and M contains an element x of order $\frac{(q+1)(q^{\frac{n}{2}-1}+1)}{\gcd(q+1,q^{\frac{n}{2}-1}+1)}$ and action type $2 \oplus (n-2)$. As $nq/2$ is odd, from Lemma 2.2, we have

$$o(x) = \frac{(q+1)(q^{\frac{n}{2}-1}+1)}{2}.$$

8.1 Orthogonal Groups of Witt Defect 0 of Dimension at Least 10

Let $x' := x^2$ and let M_0 be the subgroup of M with $|M : M_0| = 2$ and with $M_0 \cong \Omega_{n/2}(q^2)$. Observe that $x' \in M_0$. Now, x' fixes a 1-dimensional non-degenerate $\mathbb{F}_{q^2}$-subspace U of $\mathbb{F}_{q^2}^{n/2}$. Let $(M_0)_U$ be the stabilizer in M_0 of U. In particular, $x' \in (M_0)_U$. The action of x' on U induces a matrix having order $(q+1)/2$ because the element x induces by hypothesis a matrix having order $q + 1$ on U. From [66, Proposition 4.1.6], we have $(M_0)_U \cong \Omega^-_{n/2-1}(q^2).2$. Therefore, the elements of $(M_0)_U$ induce in their action on U elements having order 1 or 2. Thus, we must have

$$\frac{q+1}{2} = 2,$$

that is, $q = 3$. Now that we have pinned down the exact value of q, we use the explicit description of M in [19, Lemma 5.3.5]. Indeed, since $q = 3$, we have $q \equiv -1 \pmod{4}$ and hence [19, Lemma 5.3.5] gives $M \cong SO_{n/2}(q^2)$. As $x \in M_U$ and $M_U \cong (SO_1(q^2) \perp SO^-_{n/2-1}(q^2)).2$, we deduce that the action of x on U induces a matrix of order 2. However, this gives $q + 1 = 2$, which is clearly a contradiction. □

In order to explain the notation in the rest of this section, we recall that for n a positive integer, $[n]$ denotes an arbitrary group of order n.

Lemma 8.2 *Let M be a maximal subgroup of $\Omega^+_n(q)$ with $n \geq 10$ containing an element y of order*

$$\frac{(q^2+1)(q^{\frac{n}{2}-2}+1)}{\gcd(q^2+1, q^{\frac{n}{2}-2}+1)} = \frac{(q^2+1)(q^{\frac{n}{2}-2}+1)}{\gcd(2, q-1)}$$

and action type $4 \oplus (n - 4)$. Then, one of the following holds

(1) $M \cong (\Omega^-_4(q) \perp \Omega^-_{n-4}(q)).2^{\gcd(2,q-1)} \in \mathcal{C}_1$,
(2) $n/2$ *is even and* $M \cong \Omega^+_{n/2}(q^2).[4] \in \mathcal{C}_3$.

Note that the existence of such an element y in $\Omega^+_n(q)$ is guaranteed by Proposition 2.8.

Proof When $(n, q) = (10, 2)$, an inspection in [29] shows that $M = (\Omega^-_4(2) \perp \Omega^-_6(2)).2$. Assume then that $(n, q) \neq (10, 2)$. Then $P_{n-4}(q) \neq \emptyset$ and y is a strong $ppd(n, q; n - 4)$-element. By Theorem 2.6, M belongs to one of the Aschbacher classes $\mathcal{C}_i$, for $i \in \{1, 2, 3, 5\}$ or to $\mathcal{S}$ and in such case it is described in Example 2.6 a) of [53].

The Subgroup M Lies in Class $\mathcal{C}_1$ We use [66, Propositions 4.1.6, 4.1.7 and 4.1.20] for the structure of M.

The $\mathbb{F}_q\langle y \rangle$-module V decomposes as the sum of two irreducible submodules, say U_1 and U_2. Now, $\dim_{\mathbb{F}_q} U_1 = 4$, $\dim_{\mathbb{F}_q} U_2 = n - 4$ and the quadratic form preserved by $\Omega^+_n(q)$ restricted to both U_1 and U_2 is non-degenerate with Witt defect 1. Using this information and [66], it is readily seen that $M \cong (\Omega_4^-(q) \perp \Omega_{n-4}^-(q)).2^{\gcd(2,q-1)}$ and part (1) holds.

The Subgroup M Lies in Class C_2 Here M is described in Example 2.3 of [53]. Thus

$$M \le O_1(q) \text{ wr } S_n.$$

Let $r \in P_{n-4}(q)$ and $s \in P_4(q)$ and observe that $n-4 \ne 4$ implies $r \ne s$. Moreover, we have $\gcd(rs, |O_1(q)|) = 1$. As rs divides $o(y)$, we deduce that S_n contains an element of order rs. Therefore, from (2.4) and from the fact that r and s are distinct primes, we deduce $n \ge r + s > (n-4) + 4 = n$, which is a contradiction.

The Subgroup M Lies in Class C_3 We use [66, Propositions 4.3.14, 4.3.18 and 4.3.20] for the structure of M. One of the following holds

- $n/2$ is even and M is of type $GU_{n/2}(q)$,
- M is of type $O_{n/s}^+(q^s)$ for some prime divisor s of n, n/s even and $n/s \ge 4$,
- $qn/2$ is odd and M is of type $O_{n/2}(q^2)$.

We show that the first possibility cannot arise. Indeed pick $r \in P_{n-4}(q)$ and note that $r \ne p$, where p is the characteristic of $\mathbb{F}_q$. As r divides $o(y)$, r divides also $|M|$. But r cannot divide the factors of $|M|$ of type $q^i - 1$ for i even with $i \in \{2, \ldots, n/2\}$ because, since $n \ge 10$, we have $n/2 < n - 4$. Moreover, r cannot divide the factors of $|M|$ of type $q^i + 1$ for i odd with $i \in \{1, \ldots, n/2 - 1\}$ because this implies $n - 4 \mid 2i$. But since $n \equiv 0 \pmod{4}$, we have that $4 \mid n - 4$ and thus $2 \mid i$, against i odd.

Also the third possibility does not arise. Indeed, if M is of type $O_{n/2}(q^2)$ with $n/2$ odd, then

$$\pi(|M|) = \pi \left(p \cdot \prod_{i=1}^{\frac{n-2}{4}} (q^{4i} - 1) \right).$$

Let $r \in P_{n-4}(q)$. As before $r \ne p$ and r divides $|M|$. Hence $n - 4$ divides $4i$, for some $i \in \{1, \ldots, (n-2)/4\}$. Now $n \equiv 2 \pmod{4}$, so that also $n - 4 \equiv 2 \pmod{4}$. It follows that $n - 4 \mid 2i$, which implies $n - 4 \le (n-2)/2$ and hence $n \le 6$. However, this is a contradiction.

Finally, suppose M is of type $O_{n/s}^+(q^s)$ for some prime divisor s of n, n/s even and $n/s \ge 4$. Then

$$\pi(|M|) = \pi \left(s \cdot p \cdot (q^{\frac{n}{2}} - 1) \cdot \prod_{i=1}^{\frac{n}{2s}-1} (q^{2is} - 1) \right).$$

8.1 Orthogonal Groups of Witt Defect 0 of Dimension at Least 10   117

Pick $r \in P_{n-4}(q)$. By $n \geq 10$ and by (2.4), $r \nmid s \, p(q^{\frac{n}{2}} - 1)$. If r divides $q^{2is} - 1$ for some $i \in \{1, \ldots, n/(2s) - 1\}$ then $n - 4 \leq 2is \leq n - 2s$ and hence $s = 2$. In particular, $n/2$ is even and $M = \Omega^+_{n/2}(q^2).[4]$, that is, part (2) is satisfied.

The Subgroup M Lies in Class C_5 This case is ruled out by Lemma 2.7 (3).

The Subgroup M Lies in Class S Here M is described in Example 2.6 a) of [53] and thus

$$A_m \leq M \leq S_m \times \mathbf{Z}(\Omega^+_n(q)),$$

with $m \in \{n + 1, n + 2\}$. Thus the symmetric group S_m contains an element having order

$$\frac{\mathbf{o}(y)}{\gcd(2, q - 1)} = \frac{(q^2 + 1)(q^{\frac{n}{2}-2} + 1)}{\gcd(2, q - 1)^2}.$$

Arguing as in the symplectic case, we deduce

$$\log\left(\frac{(q^2 + 1)(q^{\frac{n}{2}-2} + 1)}{\gcd(2, q - 1)^2}\right) \leq \sqrt{m \log m}\left(1 + \frac{\log(\log(m)) - 0.975}{2\log(m)}\right).$$

For $n \geq 10$, this inequality holds true only when $q = 2$ and $n \leq 32$, or $q = 3$ and $n \leq 12$. For these cases, we have computed explicitly $\mathbf{o}(y)$ and the order of the elements of S_{n+2} and we have verified that in no case $\frac{\mathbf{o}(y)}{\gcd(2,q-1)}$ is the order of a permutation in S_{n+2}.

$\square$

Proposition 8.3 *For every $n \geq 10$, the weak normal covering number of $P\Omega^+_n(q)$ is at least 3.*

Proof As $n \neq 8$, we may argue with $\Omega^+_n(q)$. When $(n, q) = (14, 2)$, the veracity of the statement is confirmed with a computer computation; therefore, for the rest of our argument, we suppose $(n, q) \neq (14, 2)$.

Let $\tilde{H}$ be a maximal component of a weak normal covering of $\Omega^+_n(q)$ containing an $\mathrm{Aut}(\Omega^+_n(q))$-conjugate of the element x as in Lemma 8.1. Similarly, let $\tilde{K}$ be a maximal component of a weak normal covering of $\Omega^+_n(q)$ containing an $\mathrm{Aut}(\Omega^+_n(q))$-conjugate of the element y as in Lemma 8.2. It is readily seen in those lemmas that $\tilde{H}$ and $\tilde{K}$ are not $\mathrm{Aut}(\Omega^+_n(q))$-conjugate and hence $\gamma_w(\Omega^+_n(q)) \geq 2$. Assume now, by contradiction, that $\gamma_w(\Omega^+_n(q)) = 2$. Thus a weak normal 2-covering for $\Omega^+_n(q)$ is given by $\tilde{\mu} = \{\tilde{H}, \tilde{K}\}$, where $\tilde{H}$ is one of the maximal subgroups appearing in Lemma 8.1 and $\tilde{K}$ is one of the maximal subgroups appearing in Lemma 8.2.

Let $g \in \Omega^+_n(q)$ with $\mathbf{o}(g) = \frac{q^{\frac{n}{2}-1}-1}{\gcd(2,q-1)}$ and having type $(n/2 - 1) \oplus (n/2 - 1) \oplus 1 \oplus 1$. This element g arises from the embedding of $(\mathrm{O}^+_{n-2}(q) \perp \mathrm{O}^+_2(q)) \cap \Omega^+_n(q)$

in $\Omega_n^+(q)$ and hence $V = U_1 \oplus U_2 \oplus U_3 \oplus U_4$, where U_1, U_2, U_3, U_4 are irreducible $\mathbb{F}_q\langle g\rangle$-modules, $\dim_{\mathbb{F}_q} U_1 = \dim_{\mathbb{F}_q} U_2 = n/2 - 1$, $\dim_{\mathbb{F}_q} U_3 = \dim_{\mathbb{F}_q} U_4 = 1$, U_1, U_2, U_3, U_4 are totally isotropic, $U_1 \oplus U_2$ is non-degenerate and the orthogonal form preserved by $\Omega_n^+(q)$ restricted to $U_1 \oplus U_2$ is of Witt defect 0.

Let r be a primitive prime divisor of $q^{n/2-1} - 1$: the existence of r is guaranteed by Zsigmondy's theorem and by the fact that $(n, q) \neq (14, 2)$. Observe that r does not divide the order of $(\Omega_4^-(q) \perp \Omega_{n-4}^-(q)).2^{\gcd(2,q-1)}$ and of $\Omega_{n/2}^+(q^2).[4]$ when $n/2$ is even. Therefore, g is not $\mathrm{Aut}(\Omega_n^+(q))$-conjugate to an element of $\tilde{K}$.

When $n/2$ is even, r is also relatively prime to the order of $\mathrm{GU}_{n/2}(q).2$ and hence $\tilde{H}$ is not as in part (2) of Lemma 8.1.

The only 2-dimensional subspace of $\mathbb{F}_q^n$ left invariant by g is $U_1 \oplus U_2$. Since the quadratic form induced on $U_1 \oplus U_2$ has Witt defect 0, we deduce that $\tilde{H}$ is not as in part (1) of Lemma 8.1. Therefore, $nq/2$ is odd and $\tilde{H}$ is of type $\Omega_{n/2}(q^2).2 \in \mathcal{C}_3$. From Lemma 8.2, we deduce that $\tilde{K} \cong (\Omega_4^-(q) \perp \Omega_{n-4}^-(q)).2^2$.

By [90, Case (C), p. 38], $\Omega_n^+(q)$ contains a unipotent element u having Jordan blocks of size $n-1$ and 1. Suppose that u has an $\mathrm{Aut}(\Omega_n^+(q))$-conjugate in $\tilde{H}$. Since q is odd, $\Omega_{n/2}(q^2)$ contains a unipotent element having Jordan blocks of size $n-1$ and 1. Let the Jordan blocks of u, viewed as a matrix of $\mathrm{GL}_{n/2}(\mathbb{F}_{q^2})$ have sizes $\ell_1, \ell_2, \ldots, \ell_s$. Then the Jordan blocks of u, viewed as a matrix of $\mathrm{GL}_n(\mathbb{F}_q)$ have sizes $\ell_1, \ell_1, \ell_2, \ell_2, \ldots, \ell_s, \ell_s$. However, this contradicts the fact that the Jordan blocks of u, viewed as a matrix of $\mathrm{GL}_n(\mathbb{F}_q)$ have sizes $n-1$ and 1. Therefore, u has an $\mathrm{Aut}(\Omega_n^+(q))$-conjugate in $\tilde{K} \cong (\Omega_4^-(q) \perp \Omega_{n-4}^-(q)).2^2$. However, this is incompatible with the Jordan blocks of u. □

It remains to deal with the eight dimensional orthogonal group $\mathrm{P}\Omega_8^+(q)$.

8.2 Eight Dimensional Orthogonal Groups with Witt Defect 0

We recall that we have to take extra care in this case. In particular, the action of the automorphism group of $\mathrm{P}\Omega_8^+(q)$ on $\mathrm{P}\Omega_8^+(q)$ does not preserve the Jordan form of its elements. Thus we will be mainly arguing with $\mathrm{P}\Omega_8^+(q)$ and with element orders, rather than with $\Omega_8^+(q)$ and with action types. Nevertheless, it is more convenient to give the structure of maximal subgroups of $\Omega_8^+(q)$, rather than of their projective image in $\mathrm{P}\Omega_8^+(q)$.

From the discussion at the beginning of the section, we have that, there exist semisimple elements $a, b \in \mathrm{SO}_8^+(q)$ such that

- $s := ab \in \Omega_8^+(q)$,
- $\mathbf{o}(s) = \mathbf{o}(a) = \mathbf{o}(b) = q^2 + 1$,
- s has action type $4 \oplus 4$,
- $\dim_{\mathbb{F}_q} \mathbf{C}_V(a) = \dim_{\mathbb{F}_q} \mathbf{C}_V(b) = 4$, $V = \mathbf{C}_V(a) \perp \mathbf{C}_V(b)$ and

8.2 Eight Dimensional Orthogonal Groups with Witt Defect 0

- the quadratic form preserved by $\Omega_8^+(q)$ restricted to $\mathbf{C}_V(a)$ and $\mathbf{C}_V(b)$ is non-degenerate and of Witt defect 1.

The projective image $\bar{s}$ of s in $\mathrm{P}\Omega_8^+(q)$ has order $(q^2+1)/\gcd(2, q-1)$, because $\langle s \rangle$ contains the matrix $-I$, generator of $\mathbf{Z}(\Omega_8^+(q))$.

We claim that, when $q > 3$, we may choose a and b such that a and b are not in the same $\Omega_8^+(q)$-conjugacy class. The proof is very similar to the proof of Lemma 5.3, here we only give a sketch of the proof. Indeed, $|\mathbf{N}_{\Omega_8^+(q)}(\langle a \rangle) : \mathbf{C}_{\Omega_8^+(q)}(\langle a \rangle)| = 4$ and the action of $\mathbf{N}_{\Omega_8^+(q)}(\langle a \rangle)$ on $\langle a \rangle$ is the cyclic action of the Galois group of $\mathbb{F}_{q^4}/\mathbb{F}_q$. In particular, as long as $4 < \phi(o(a))$, we may always find b not in the same $\Omega_8^+(q)$-conjugacy class of a. It follows from an easy computation that, if $\phi(x) \leq 4$, then $x \leq 12$. Note now that the only numbers $q^2 + 1$ that are less than or equal to 12 arise when $q \in \{2, 3\}$.

Suppose now that $q > 3$ and let s, a, b be as in the previous paragraphs and with a and b not in the same $\Omega_8^+(q)$-conjugacy class. A computation yields that $\mathbf{C}_{\mathrm{P}\Omega_8^+(q)}(\bar{s})$ is contained in the projective image of $(\mathrm{O}_4^-(q) \perp \mathrm{O}_4^-(q)) \cap \Omega_8^+(q)$ in $\mathrm{P}\Omega_8^+(q)$. Hence

$$\mathbf{C}_{\mathrm{P}\Omega_8^+(q)}(\bar{s}) = \frac{(\langle a \rangle \times \langle b \rangle) \cap \Omega_8^+(q)}{\mathbf{Z}(\Omega_8^+(q))}$$

and

$$|\mathbf{C}_{\mathrm{P}\Omega_8^+(q)}(\bar{s})| = \frac{(q^2+1)^2}{\gcd(2, q-1)} = \begin{cases} (q^2+1)^2 & \text{when } q \text{ is even,} \\ \frac{(q^2+1)^2}{2} & \text{when } q \text{ is odd.} \end{cases}$$

Using [9, Table 8.50], when $q > 3$, we have selected in Table 8.1 the maximal subgroups $\tilde{H}$ of $\Omega_8^+(q)$ such that $H \leq \mathrm{P}\Omega_8^+(q)$ contains a semisimple element $\bar{s}$ having order $(q^2+1)/\gcd(2, q-1)$ (regardless of the action of its lift s to $\Omega_8^+(q)$ on V).

Using Table 8.1, it can be shown that the only maximal subgroups $\tilde{H}$ of $\Omega_8^+(q)$ such that $H \leq \mathrm{P}\Omega_8^+(q)$ contains a semisimple element $\bar{s}$ with $o(\bar{s}) = (q^2+1)/\gcd(2, q-1)$ and $|\mathbf{C}_{\mathrm{P}\Omega_8^+(q)}(\bar{s})| = (q^2+1)^2/\gcd(2, q-1)$ are given in Table 8.2.

Lemma 8.4 *The weak normal covering number of $\mathrm{P}\Omega_8^+(q)$ is at least 2. Moreover, if H and K are maximal subgroups of a weak normal 2-covering of $\mathrm{P}\Omega_8^+(q)$, then $q \in \{2, 3\}$ and, up to $\mathrm{Aut}(\mathrm{P}\Omega_8^+(q))$-conjugacy, H and K are one of the examples in Table 1.7.*

Proof When $q \in \{2, 3, 5\}$, we have verified the veracity of this lemma with a computer. Therefore, for the rest of the proof we suppose $q \notin \{2, 3, 5\}$.

Let H be a maximal component of a weak normal 2-covering of $\mathrm{P}\Omega_8^+(q)$ containing an element having order $(q^2+1)/\gcd(2, q-1)$ and having centralizer of order $(q^2+1)^2/\gcd(2, q-1)$. Then $\tilde{H}$ is described in Table 8.2. In particular, the order of H is given in Table 8.3.

Table 8.1 Maximal subgroups $\tilde{H}$ of $\Omega_8^+(q)$ such that H contains a semisimple element of order $(q^2+1)/\gcd(2, q-1)$

q odd, $q > 3$				q even, $q > 2$		
Structure	Class	nr. $\Omega_8^+(q)$-classes		Structure	Class	nr. $\Omega_8^+(q)$-classes
$E_q^6 : \left(\frac{q-1}{2} \times \Omega_6^+(q)\right).2$	$\mathcal{C}_1$	1		$E_q^6 : ((q-1) \times \Omega_6^+(q))$	$\mathcal{C}_1$	1
$E_q^6 : \frac{1}{2}\mathrm{GL}_4(q)$	$\mathcal{C}_1$	2		$E_q^6 : \mathrm{GL}_4(q)$	$\mathcal{C}_1$	2
$2 \times \Omega_7(q)$	$\mathcal{C}_1$	2		$\mathrm{Sp}_6(q)$	$\mathcal{C}_1$	1
$2\cdot\Omega_7(q)$	$\mathcal{S}$	4		$\mathrm{Sp}_6(q)$	$\mathcal{S}$	2
$(\Omega_2^+(q) \times \Omega_6^+(q)).[4]$	$\mathcal{C}_1$	1		$(\Omega_2^+(q) \times \Omega_6^+(q)).2$	$\mathcal{C}_1$	1
$\mathrm{SL}_4(q).\frac{q-1}{2}.2$	$\mathcal{C}_2$	2		$\mathrm{SL}_4(q).(q-1).2$	$\mathcal{C}_2$	2
$(\Omega_2^-(q) \times \Omega_6^-(q)).[4]$	$\mathcal{C}_1$	1		$(\Omega_2^-(q) \times \Omega_6^-(q)).2$	$\mathcal{C}_1$	1
$\mathrm{SU}_4(q).\frac{q+1}{2}.2$	$\mathcal{C}_3$	2		$\mathrm{SU}_4(q).(q+1).2$	$\mathcal{C}_3$	2
$\Omega_4^-(q)^2.[4].S_2$	$\mathcal{C}_2$	1		$\Omega_4^-(q)^2.2.S_2$	$\mathcal{C}_2$	1
$\Omega_4^+(q^2).[4]$	$\mathcal{C}_3$	2		$\Omega_4^+(q^2).[4]$	$\mathcal{C}_3$	2
$2 \times \Omega_8^-(\sqrt{q})$	$\mathcal{C}_5$	2		$\Omega_8^-(\sqrt{q})$	$\mathcal{C}_5$	1
$2\cdot\Omega_8^-(\sqrt{q})$	$\mathcal{S}$	4		$\Omega_8^-(\sqrt{q})$	$\mathcal{S}$	2
$(\Omega_3(q) \times \Omega_5(q)).[4]$	$\mathcal{C}_1$	2				
$(\mathrm{Sp}_2(q) \circ \mathrm{Sp}_4(q)).2$	$\mathcal{C}_4$	4				
$q = 5, 2\cdot\mathrm{Suz}(8)$	$\mathcal{S}$	8				

Table 8.2 Maximal subgroups $\tilde{H}$ of $\Omega_8^+(q)$ such that H contains a semisimple element of order $(q^2+1)/\gcd(2, q-1)$ and having small centralizer

q odd, $q > 3$			q even, $q > 2$		
Structure	Class	nr. $\Omega_8^+(q)$-cl.s	Structure	Class	nr. $\Omega_8^+(q)$-cl.s
$\Omega_4^-(q)^2.[4].S_2$	$\mathcal{C}_2$	1	$\Omega_4^-(q)^2.2.S_2$	$\mathcal{C}_2$	1
$\Omega_4^+(q^2).[4]$	$\mathcal{C}_3$	2	$\Omega_4^+(q^2).[4]$	$\mathcal{C}_3$	2
$2 \times \Omega_8^-(\sqrt{q})$	$\mathcal{C}_5$	2	$\Omega_8^-(\sqrt{q})$	$\mathcal{C}_5$	1
$2\cdot\Omega_8^-(\sqrt{q})$	$\mathcal{S}$	4	$\Omega_8^-(\sqrt{q})$	$\mathcal{S}$	2
$q = 5, 2\cdot\mathrm{Suz}(8)$	$\mathcal{S}$	8			

Table 8.3 Possible orders of H

q odd, $q > 3$		q even, $q > 2$	
Order H	Comments	Order H	Comments
$2q^4(q^4-1)^2$		$4q^4(q^4-1)^2$	
$q^6(q-1)(q^3-1)(q^4+1)$	q a square	$q^6(q-1)(q^3-1)(q^4+1)$	q a square
29120	$q = 5$		

From the embedding of $O_6^-(q) \perp O_2^-(q)$ in $O_8^+(q)$, using Proposition 2.8, we deduce that $\Omega_8^+(q)$ contains a semisimple element s_- having order $q^3 + 1$ and $P\Omega_8^+(q)$ contains an element $\bar{s}_-$ having order $(q^3 + 1)/\gcd(2, q - 1)$. Similarly,

8.2 Eight Dimensional Orthogonal Groups with Witt Defect 0

Table 8.4 Maximal subgroups of order divisible by $(q^6 - 1)/\gcd(2, q - 1)$

q odd, $q > 3$			q even, $q > 2$		
Structure	Class	nr. $O_8^+(q)$-classes	Structure	Class	nr. $O_8^+(q)$-classes
$2 \times \Omega_7(q)$	$\mathcal{C}_1$	2	$\mathrm{Sp}_6(q)$	$\mathcal{C}_1$	1
$2^{\cdot}\Omega_7(q)$	$\mathcal{S}$	4	$\mathrm{Sp}_6(q)$	$\mathcal{S}$	2

from the embedding of $O_6^+(q) \perp O_2^+(q)$ in $O_8^+(q)$, using Lemma 7.1, we deduce that $\Omega_8^+(q)$ contains a semisimple element s_+ having order $q^3 - 1$ and $P\Omega_8^+(q)$ contains an element $\bar{s}_+$ having order $(q^3 - 1)/\gcd(2, q - 1)$. It is readily seen from Table 8.3 that $|H|$ is not divisible by a primitive prime divisor of $q^6 - 1$ and hence H does not contain an $\mathrm{Aut}(P\Omega_8^+(q))$-conjugate of $\bar{s}_-$. Observe now that, if H contains an $\mathrm{Aut}(P\Omega_8^+(q))$-conjugate of $\bar{s}_+$, then $(q^3 - 1)/\gcd(2, q - 1)$ divides $|H|$ and hence $\tilde{H}$ is isomorphic to either $2 \cdot \Omega_8^-(\sqrt{q})$ or $2 \times \Omega_8^-(\sqrt{q})$ when q is odd, or to $\Omega_8^-(\sqrt{q})$ when q is even. However, these groups do not contain elements having order as large as $(q^3 - 1)/\gcd(2, q - 1) = (\sqrt{q}^6 - 1)/\gcd(2, \sqrt{q} - 1)$. Therefore, H does not contain an $\mathrm{Aut}(P\Omega_8^+(q))$-conjugate of $\bar{s}_+$.

The previous paragraph shows that $\gamma_w(P\Omega_8^+(q)) \geq 2$. We now argue by contradiction and we suppose that $\gamma_w(P\Omega_8^+(q)) = 2$. Let K be the second component in a weak normal 2-covering of $P\Omega_8^+(q)$, with K maximal. By the above analysis, we get that $\tilde{K}$ contains elements having order $q^3 - 1$ and $q^3 + 1$. Therefore, by Lemma 2.2, $|\tilde{K}|$ is divisible by

$$\frac{(q^3 - 1)(q^3 + 1)}{\gcd(q^3 - 1, q^3 + 1)} = \frac{q^6 - 1}{\gcd(2, q - 1)}.$$

A direct inspection on the maximal subgroups of $\Omega_8^+(q)$ in [9, Table 8.50] reveals that $\tilde{K}$ is one of the groups in Table 8.4.

Suppose first that q is even. In particular, $\Omega_8^+(q) = P\Omega_8^+(q)$, $H = \tilde{H}$ and $K = \tilde{K} \cong \mathrm{Sp}_6(q)$. Observe that the triality automorphism of $\Omega_8^+(q)$ fuses the three $\Omega_8^+(q)$-classes of maximal subgroups isomorphic to $\mathrm{Sp}_6(q)$. We now consider particular elements of order $q^3 + 1$. We consider the elements g of order $q^3 + 1$ and with the property that V has a 6-dimensional irreducible $\mathbb{F}_q\langle g\rangle$-submodule W with g inducing on W a matrix of order $q^3 + 1$. These elements fall into two types: the elements having two eigenvalues in $\mathbb{F}_q$ (and hence having type $6 \oplus 1 \oplus 1$ on V) and the elements having no eigenvalues in $\mathbb{F}_q$ (and hence having type $6 \oplus 2$ on V). The elements of the first type are covered by the $\Omega_8^+(q)$-class of $\mathrm{Sp}_6(q)$ in class $\mathcal{C}_1$. Therefore the remaining type is covered by the remaining two $\Omega_8^+(q)$-classes of $\mathrm{Sp}_6(q)$ in class $\mathcal{S}$. Let us denote by o_1 the number of elements of the first type and by o_2 the number of elements of the second type. We claim that

$$o_2 = o_1 q.$$

Indeed, any element g of the first type decomposes V as the direct sum $V = W \oplus W^\perp$, where $\dim_{\mathbb{F}_q}(W) = 6$, $\dim_{\mathbb{F}_q}(W^\perp) = 2$ and the quadratic forms induced on W and on $W^\perp$ are of Witt defect 1. As $\Omega_2^-(q)$ has order $q+1$, we can use the same g to construct $(q+1) - 1 = q$ distinct elements of the second type. This proves our claim.

As the $\Omega_8^+(q)$-class of $\mathrm{Sp}_6(q) \in \mathcal{C}_1$ covers o_1 elements of order $q^3 + 1$, the remaining two classes cover at most $2o_1$ elements of order $q^3 + 1$. Observe that here we are using the fact that the triality automorphism of $\Omega_8^+(q)$ fuses the three $\Omega_8^+(q)$-classes of maximal subgroups isomorphic to $\mathrm{Sp}_6(q)$. Therefore, these two classes cover all elements of the second type only if

$$2o_1 \geq o_2 = o_1 q,$$

that is, $q \leq 2$, which is a contradiction. This argument explain also why the case $q = 2$ is special and had to be dealt with by a computer.

Suppose now that q is odd. The argument above can be applied also when q is odd, the only difference is that $q + 1$ is replaced by $(q + 1)/2$. Therefore we have a putative weak normal covering only if $2 \geq (q - 1)/2$, that is, $q \leq 5$. As we mentioned in the opening paragraph of this proof, the cases $q \in \{3, 5\}$ have been dealt with by a computer. □

The proof of Lemma 8.4 is necessarily technical because the weak normal covering number of $\mathrm{P}\Omega_8^+(q)$ is not that far from 2. Indeed, $\mathrm{P}\Omega_8^+(q)$ has a weak normal covering having three components, see Theorem 8.6. Therefore, in view of Lemma 8.4, $\gamma_w(\mathrm{P}\Omega_8^+(q)) = 3$ when $q \geq 4$ and $\gamma_w(\mathrm{P}\Omega_8^+(q)) = 2$ when $q \leq 3$. To prove Theorem 8.6, we first need a preliminary observation concerning $\Omega_4^+(q)$.

Lemma 8.5 *Let $\tilde{g} \in \Omega_4^+(q)$ having order divisible by the characteristic p of $\mathbb{F}_q$. Then $\tilde{g}$ in its action on the orthogonal space $\mathbb{F}_q^4$ fixes a 2-dimensional totally isotropic subspace.*

Proof Let $G := \mathrm{SL}_2(q) \times \mathrm{SL}_2(q)$ and let V be the 4-dimensional vector space over $\mathbb{F}_q$ consisting of the 2×2 matrices with coefficients in $\mathbb{F}_q$. The group G has a linear action on V by setting

$$(a, b) \cdot v := a^{-1} v b,$$

for each $(a, b) \in G$ and $v \in V$. Let $\tilde{G}$ be the group induced by the linear action of G on V.

Now, V is endowed with a natural quadratic form $Q : V \to \mathbb{F}_q$ by setting

$$Q(v) := \det v = x_{11} x_{22} - x_{12} x_{21}, \qquad \text{for all } v = \begin{pmatrix} x_{11} & x_{12} \\ x_{21} & x_{22} \end{pmatrix} \in G.$$

In particular, $v \in V$ is totally singular if and only if $\det v = 0$.

8.2 Eight Dimensional Orthogonal Groups with Witt Defect 0

The linear action of G on V preserves the quadratic form Q because

$$Q((a,b) \cdot v) = Q(a^{-1}vb) = \det(a^{-1}vb) = \det(v) = Q(v).$$

In particular, $\tilde{G} \leq \mathrm{SO}_4^+(q)$, because Q has Witt defect zero. It is not hard to verify that $\tilde{G} \leq \Omega_4^+(q)$. Since $|\tilde{G}| = |G|/2 = q^2(q^2-1)/2 = |\Omega_4^+(q)|$, we deduce $\tilde{G} = \Omega_4^+(q)$. We use this model of $\Omega_4^+(q)$ to prove this lemma.

Let $\tilde{g} \in \tilde{G}$ be an element having order divisible by the characteristic of $\mathbb{F}_q$. In particular, $\tilde{g}$ is the projection of an element $g \in G$. Replacing g by a suitable conjugate, we may suppose that

$$g = \left(\begin{pmatrix} 1 & a \\ 0 & 1 \end{pmatrix}, b \right) \text{ or } g = \left(b, \begin{pmatrix} 1 & a \\ 0 & 1 \end{pmatrix} \right),$$

for some $a \in \mathbb{F}_q$ and some $b \in \mathrm{SL}_2(q)$. In the first case, g fixes the 2-dimensional totally isotropic subspace

$$\left\langle \begin{pmatrix} 1 & 0 \\ 0 & 0 \end{pmatrix}, \begin{pmatrix} 0 & 1 \\ 0 & 0 \end{pmatrix} \right\rangle.$$

In the second case, g fixes the 2-dimensional totally isotropic subspace

$$\left\langle \begin{pmatrix} 0 & 1 \\ 0 & 0 \end{pmatrix}, \begin{pmatrix} 0 & 0 \\ 0 & 1 \end{pmatrix} \right\rangle.$$

□

Theorem 8.6 *The group $\mathrm{P}\Omega_8^+(q)$ admits a weak normal covering having cardinality 3.*

Proof Let H be the stabilizer of a 1-dimensional totally singular subspace of $V = \mathbb{F}_q^8$, let J be the stabilizer of a 2-dimensional non-degenerate subspace of V of minus type, and let K be the stabilizer of a 4-dimensional non-degenerate subspace of $V = \mathbb{F}_q^8$ of minus type. We verify, using the triality, that H, J, K are the components of a weak normal covering of $\mathrm{P}\Omega_8^+(q)$.

Let $g \in \mathrm{P}\Omega_8^+(q)$. We show that g has an $\mathrm{Aut}(\mathrm{P}\Omega_8^+(q))$-conjugate in H, J or K. We consider the group $\Omega_8^+(q)$ and we let $\tilde{g} \in \Omega_8^+(q)$ projecting to g. Since $\Omega_8^+(q)$ admits no Singer cycle, $\tilde{g}$ does not act irreducibly on V and hence $\tilde{g}$ fixes a non-zero proper subspace W of V with $1 \leq \dim(W) \leq 4$. We choose W so that W is an irreducible $\mathbb{F}_q\langle \tilde{g}\rangle$-submodule of V. In particular, W is either a totally isotropic subspace of V, or W is a non-degenerate subspace of V.

Assume that W is a non-degenerate subspace of V. Observe that $\dim W$ is even and that the quadratic form induced by V on W has Witt defect 1, because $\tilde{g}$ acts irreducibly on W and because odd dimensional orthogonal groups $\Omega_\ell(q)$ with $\ell \geq 3$

and even dimensional orthogonal groups of plus type do not contain Singer cycles. In particular, g is $\mathrm{P}\Omega_8^+(q)$-conjugate to an element of J or K.

Assume that W is a totally isotropic subspace of V. If $\dim W = 1$, then g is $\mathrm{P}\Omega_8^+(q)$-conjugate to an element of H. Notice (see [26, Section 8.5]) that the triality automorphism of $\mathrm{P}\Omega_8^+(q)$ maps stabilizers of 1-dimensional totally isotropic subspaces of V to one of the two families of maximal totally isotropic subspaces of V. Therefore, if $\dim W = 4$, then g is $\mathrm{Aut}(\mathrm{P}\Omega_8^+(q))$-conjugate to H. Suppose $\dim W = 3$. From the polar geometry associated to V [26, Section 6.4], we see that g fixes a 4-dimensional totally isotropic subspace. In particular, also in this case, g is $\mathrm{Aut}(\mathrm{P}\Omega_8^+(q))$-conjugate to an element of H. Suppose $\dim W = 2$. Now, $\tilde{g}$ fixes $W^\perp$ and hence $\tilde{g}$ induces a linear action on the 4-dimensional vector space $W^\perp/W$ having Witt defect zero: the fact that $W^\perp/W$ has the same type as V follows from the discussion on polar spaces in [26, Section 6]. Clearly, $\tilde{g}$ is not semisimple and it does induce an element on $W^\perp/W$ having order divisible by the characteristic of $\mathbb{F}_q$: indeed, if $\tilde{g}$ induces a semisimple element on $W^\perp/W$, then $\tilde{g}$ is semisimple because $\tilde{g}$ acts on W and on $V/W^\perp$ irreducibly. Therefore, by Lemma 8.5, $\tilde{g}$ fixes a 2-dimensional totally isotropic subspace of $W^\perp/W$. Hence $\tilde{g}$ fixes a 4 dimensional totally isotropic subspace of V. In particular, arguing as above, g is $\mathrm{Aut}(\mathrm{P}\Omega_8^+(q))$-conjugate to an element of H. □

Now, the veracity of Table 1.7 follows from the results in this chapter.

Chapter 9
Proofs of the Main Theorems

Theorems 1.3, 1.5 and Corollary 1.6 follow now immediately by inspection of the Tables 1.3–1.7.

We devote the rest of the section to the proof of Theorem 1.9. We start with an elementary observation. Recall that a group Y is said to be **supersolvable** if there exists a normal series $1 = Y_0 \trianglelefteq Y_1 \trianglelefteq \cdots \trianglelefteq Y_{\ell-1} \trianglelefteq Y_\ell = Y$ such that each quotient Y_i/Y_{i-1} is cyclic.

Lemma 9.1 *Let Y be a supersolvable group, let V and Z be subgroups of Y with $V < Z$ and $Z \trianglelefteq Y$. Then*

$$\bigcup_{y \in Y} V^y \subsetneq Z.$$

Proof We prove it by induction on $|Y|$. As Y is a finite group, using [2, Lemma 2 parts (ii) and (iii)], we see that there exists a normal series

$$1 = Y_0 \trianglelefteq Y_1 \trianglelefteq \cdots \trianglelefteq Y_{\ell-1} \trianglelefteq Y_\ell = Y$$

of Y, with Y_i/Y_{i-1} cyclic of prime order for each $i \in \{1, \ldots, \ell\}$ and with $Z = Y_\kappa$, for some $\kappa \in \{1, \ldots, \ell\}$. We write $\bar{Y} := Y/Y_1$ and we use the "bar" notation for the projection of Y to $\bar{Y}$.

Now, $\bar{V} \leq \bar{Z} \trianglelefteq \bar{Y}$. Suppose first $\bar{V} \neq \bar{Z}$. As $\bar{Y}$ is a quotient of Y and as Y is supersolvable, by [84, p. 150], we deduce that $\bar{Y}$ is supersolvable. Since $|\bar{Y}| < |Y|$, our inductive hypothesis gives

$$\bigcup_{\bar{y} \in \bar{Y}} \bar{V}^{\bar{y}} \subsetneq \bar{Z}.$$

From this, it immediately follows that

$$\bigcup_{y \in Y} V^y \subsetneq Z.$$

Suppose next that $\bar{V} = \bar{Z}$. Since $Y_1 \leq Z$, we deduce $Z = VY_1$. Let p be the order of Y_1 and assume, by contradiction, that $\bigcup_{y \in Y} V^y = Z$. Since Y_1 is cyclic included in Z, there exists $y \in Y$ with $Y_1^y \leq V$. However, since $Y_1 \trianglelefteq Y$, we get $Y_1 = Y_1^y \leq V$ and thus $Z = V$, contrary to our hypothesis. □

Next, we need a rather technical result.

Lemma 9.2 *Let $U := B \times C$ be a finite group with $B \cong S_4$ and with C cyclic of order f. Let V, Z and Y be subgroups of U with $V < Z \trianglelefteq Y \leq U$ and*

$$Z = \bigcup_{y \in Y} V^y.$$

Then the following holds:

(1) $Z = K \times C_1$, *where K is the Klein subgroup of B and $C_1 \leq C$,*
(2) $V = K_0 \times C_1$, *where K_0 is a subgroup of order 2 of K.*

Proof Since $C = Z(U)$ and since $Z = \bigcup_{y \in Y} V^y$, we get

$$C \cap V = C \cap Z. \tag{9.1}$$

Now, consider the subgroups of U given by $Y_0 = YC$, $Z_0 = ZC$ and $V_0 = VC$. Observe that Y_0/C is isomorphic to a subgroup of $U/C \cong B \cong S_4$. Moreover,

$$Z_0 = \bigcup_{y \in Y_0} V_0^y.$$

Now, a direct inspection on the subgroups of S_4 reveals that either $Z_0 = V_0$, or

- $Y_0 = B_1 \times C$, where $B_1 \cong A_4$ or $B_1 \cong S_4$,
- $Z_0 = K \times C$, where K is the Klein subgroup of B, and
- $V_0 = K_0 \times C$, where K_0 is a subgroup of order 2 of K.

The first possibility yields

$$VC = V_0 = Z_0 = ZC. \tag{9.2}$$

However, (9.1) and (9.2) yield $V = Z$, which is a contradiction. Thus

$$VC = K_0 \times C \text{ and } ZC = K \times C.$$

Let $\pi : Z \to C$ be the natural projection and let C_1 be the image of π. To conclude the proof it suffices to show that $C_1 \leq V$ (indeed, this implies $Z = K \times C_1$ and $V = K_0 \times C_1$). Let $c \in C_1$ with $C_1 = \langle c \rangle$ and let k_1, k_2, k_3 be the three involutions in K. Since c lies in the image of π, there exists $z \in Z$ with $c = \pi(z)$. Now, as $Z \leq K \times C$, we have $z = k_i c$, for some $i \in \{1, 2, 3\}$. Without loss of generality, we may suppose that $i = 1$. As Z is a union of Y-conjugates, we may assume, replacing z with a suitable Y-conjugate, that $z \in V$. Let $y \in Y$ with $k_1^y = k_2$. Observe that the existence of y is guaranteed by the fact that YC/C contains a subgroup isomorphic to A_4. Now,

$$z^y = k_1^y c^y = k_2 c.$$

Thus $k_2 c \in Z$ and hence $k_3 = (k_1 c)(k_2 c)^{-1} \in Z$. In particular, $k_i \in Z$, for every $i \in \{1, 2, 3\}$. Thus $c = (k_1 c) k_1^{-1} \in Z$. Now, c has a Y-conjugate in V and hence $c \in V$. □

Proof of Theorem 1.9 Let X be an almost simple group with socle G, let $A_X := \text{Aut}(X)$ and $A_G := \text{Aut}(G)$. We have that

$$G \leq X \trianglelefteq A_X \leq A_G. \tag{9.3}$$

In particular A_X is almost simple with socle G and G is characteristic in every term of the chain (9.3).

Suppose now that part (1) of Theorem 1.9 does not hold. Then $\gamma_w(X) = 1$. Let $\mu := \{H\}$ be a weak normal 1-covering of X for some $H < X$. We need to show that $G = P\Omega_8^+(q)$, q is odd, A_X contains a triality automorphism of G, X does not contain a triality automorphism of G and $G \leq H$.

As

$$X = \bigcup_{a \in A_X} H^a, \tag{9.4}$$

we get

$$G = G \cap \bigcup_{a \in A_X} H^a = \bigcup_{a \in A_X} (G \cap H)^a \subseteq \bigcup_{a \in A_G} (G \cap H)^a \subseteq G$$

and hence

$$G = \bigcup_{a \in A_G} (G \cap H)^a. \tag{9.5}$$

Assume $G \not\leq H$, that is, $G \cap H$ is a proper subgroup of G. From (9.5), $\{G \cap H\}$ is a weak normal 1-covering of G and $\gamma_w(G) = 1$, contradicting Theorem 1.3. Therefore

$$G \leq H.$$

Set $Y := A_X/G$, $Z := X/G$ and $V := H/G$ and observe that $V < Z \trianglelefteq Y$. Since G is characteristic in A_X and thus also normal in A_X, we obtain that $G^a = G$, for each $a \in A_X$. Thus, the elements $a \in A_X$ induce a group automorphism on the quotient $X/G = Z$ by setting $(xG)^a = x^aG$, for every $x \in X$. In particular, we have a group action of A_X on $X/G = Z$, that is, a group homomorphism $A_X \to \mathrm{Aut}(Z)$. Observe that G is contained in the kernel of this group homomorphism and hence we have a group action of $A_X/G = Y$ on Z by setting $(xG)^{aG} = x^aG$, for every $x \in X$ and for every $a \in A_X$. As $(xG)^{aG} = (xG)^a$ for every $x \in X$ and for every $a \in A_X$, recalling that $G \leq X$ and $G \leq H$ and by quotienting the left and the right hand side of (9.4) with G, we obtain

$$Z = \frac{X}{G} = \bigcup_{a \in A_X} \left(\frac{H}{G}\right)^a = \bigcup_{a \in A_X} \left(\frac{H}{G}\right)^{aG} = \bigcup_{a \in Y} V^a.$$

Now, Lemma 9.1 shows that Y is not supersolvable. Since subgroups of supersolvable groups are supersolvable (see [84, p. 150]), we deduce that A_G is not supersolvable. A direct inspection on the outer automorphisms of the non-abelian simple groups reveals that $G = \mathrm{P}\Omega_8^+(q)$ and q is odd. We postpone the proof that $A_X = \mathrm{Aut}(X)$ contains a triality automorphism, that X does not contain a triality automorphism and that $\mathrm{PO}_8^+(q) \leq X$ to Remark 9.3, where we give slightly more information on X and H. □

Remark 9.3 Let X be an almost simple group with socle G and let $A_X := \mathrm{Aut}(X)$. Suppose $\gamma_w(X) = 1$ and let $\mu := \{H\}$ be a weak normal 1-covering of X.

Using Lemma 9.2 and the notation in the proof above, we can give much more detailed information on X, G and H arising in part (2) of Theorem 1.9.

From [66, p. 38], we have

$$\mathrm{Out}(G) = \mathrm{Out}(\mathrm{P}\Omega_8^+(q)) = B \times C,$$

where q is odd, $B \cong S_4$ is the subgroup generated by the inner-diagonal and graph outer-automorphisms of G and C is cyclic of order f and is the subgroup generated by the field outer-automorphisms.

Set $U := B \times C$ and observe that $V < Z \trianglelefteq Y \leq U$, where $V = H/G$, $Z = X/G$ and $Y = A_X/G$. In particular, Lemma 9.2 yields very detailed information on the various possibilities for Y, Z and V, and hence in turn for A_X, X and H. From this, using [49, Chapter 4], it immediately follows that A_X contains a triality automorphism, X does not contain a triality automorphism and H contains an A_X-conjugate of $\mathrm{PO}_8^+(q)$. As the A_X-conjugates of H cover X, we deduce $\mathrm{PO}_8^+(q) \leq X$.

Chapter 10
Almost Simple Groups Having Socle a Sporadic Simple Group

In this chapter we prove the correctness of Table 1.2. Our proof is mainly computational.

Let G be an almost simple group having socle a sporadic simple group. Except for the Baby Monster and for the Monster, the computations are straightforward because in each case the character table and the list of the maximal subgroups of G are stored in GAP and in magma. Therefore, for these groups, we may compute the list of all the permutation characters of the faithful primitive actions of G. Using this list, $\gamma(G)$ is simply the minimum number of permutation characters $\pi_1, \ldots, \pi_\ell$ such that the class function $\pi_1 + \cdots + \pi_\ell$ does not vanish in any conjugacy class.[1] Similarly, using this list, we may easily determine the edges of the invariably generating graph $\Lambda(G)$. Indeed, given two conjugacy classes C_1 and C_2 of G, $\{C_1, C_2\}$ is an edge of $\Lambda(G)$ if and only if, for some $c_1 \in C_1$ and $c_2 \in C_2$,

$$\pi(c_1)\pi(c_2) = 0,$$

for every permutation character π as above.

Now, with $\Lambda(G)$, we can use standard algorithms to compute the clique number $\kappa(G)$.

The computation for the Baby Monster[2] F_{2+} is slightly more involved. All the permutation characters of F_{2+} acting on the right cosets of a maximal subgroup are stored in GAP, except for the action of F_{2+} on the cosets of a maximal subgroup M of type $(2^2 \times F_4(2)) : 2$. This omission from the GAP library is due to the fact that the

---

[1] We have already used this idea in the last paragraph of the proof of Lemma 5.4.
[2] We employ the ATLAS notation F_{2+} for the Baby Monster and F_1 for the Monster group. However, adopting the friendlier notations B and M would clash with our conventional usage: typically, M denotes a maximal subgroup, while B, in subsequent discussions, will represent a specific conjugacy class within F_1.

conjugacy fusion of some of the elements of M in F_{2+} remains a mystery. Therefore, except for the action of F_{2+} on the right cosets of $(2^2 \times F_4(2)) : 2$, we may compute all the remaining permutation characters of primitive faithful actions. We have computed the possible permutation character of the action of F_{2+} on the right cosets of $(2^2 \times F_4(2)) : 2$ using the GAP command `PossibleClassFusions` as instructed in [22, Proposition 3.4]. Despite the ambiguity on the fusion of the conjugacy classes of $(2^2 \times F_4(2)) : 2$ in F_{2+}, the command `PossibleClassFusions` shows that all of these possibilities give rise to the same permutation character. Now, with all of these permutation characters available, we may argue as in all other cases.

We now deal with the case that $G = F_1$ is the Monster group. From [92, Section 3.6] and [93], we see that the classification, up to isomorphism and up to conjugacy,[3] of the maximal subgroups of G is complete except for a few open cases. In particular, if M is a maximal subgroup of G, then either M is in [92, Section 3.6], or the socle of M is $PSL_2(13)$ or $PSL_2(16)$. However, things are not that simple because the conjugacy classes of the known maximal subgroups of G are not yet fully understood. Moreover, to the best of our knowledge, we currently know only six permutation characters of primitive faithful actions of G: namely, the permutation characters for the actions of G on the right cosets of maximal subgroups of type $2.F_{2+}$, $2^{1+24}.Co_1$, $3.Fi_{24}$, $2^2.{}^2E_6(2) : S_3$ and $2^{5+10+20}.(S_3 \times L_5(2))$ are determined in [10, 11], and of type $3^{1+12}.2.Suz.2$ is determined in [4]. Therefore, for the Monster G, we have argued differently.

Given a positive integer x, we let $\delta(x)$ be the set of divisors of x and, given a set of positive integers X, we let $\Delta(X) = \bigcup_{x \in X} \delta(x)$. In Table 10.1, we have collected some information on the maximal subgroups of G. In the first column we have given the name of the maximal subgroups of G (up to the small ambiguities explained above) and in the second column we have given a set of integers X such that $\Delta(X)$ is the set of element orders of the corresponding maximal subgroup. For instance, in the first row of Table 10.1, we see that the second column is

$$\{32, 36, 38, 40, 44, 46, 48, 50, 54, 56, 60, 62, 66, 68, 70, 78, 84, 94, 104, 110\}$$

and hence the order of the elements in $2.B$ are all of the divisors of the elements in this set. The symbol $\Delta(X)$ help us to keep the notation short, so that we do not have to list all of the element orders of a maximal subgroup, but only some. In the last two lines, we have put a asterisk next to the groups $L_2(13).2$ and $L_2(16).4$ because at the moment it is not known[4] whether the Monster has maximal subgroups having socle $L_2(13)$ or $L_2(16)$. Moreover, in case that the Monster does contain such maximal subgroups, it is not clear what is the exact structure and the number of conjugacy classes. This ambiguity will play no role in our argument that follows.

---

[3] The recent paper [32] has provided a comprehensive classification of the maximal subgroups of the Monster group, although the classification is not yet complete up to conjugacy.

[4] From [32], we see that F_1 has no maximal subgroup having socle $PSL_2(16)$, but admits maximal subgroups having socle $PSL_2(13)$.

10 Almost Simple Groups Having Socle a Sporadic Simple Group

Table 10.1 Maximal subgroups of the Monster and their element orders

Subgroup	Element orders
$2.F_{2+}$	{32, 36, 38, 40, 44, 46, 48, 50, 54, 56, 60, 62, 66, 68, 70, 78, 84, 94, 104, 110}
$2^{1+24}.Co_1$	{32, 36, 40, 48, 52, 56, 60, 66, 70, 78, 84, 88, 92}
$3.Fi_{24}$	{34, 36, 40, 45, 46, 48, 51, 54, 60, 66, 69, 70, 78, 84, 87, 105}
$2^2.{}^2E_6(2):S_3$	{32, 36, 38, 40, 48, 52, 56, 60, 66, 68, 70, 84}
$2^{10+16}.O_{10}^+(2)$	{32, 34, 36, 40, 45, 48, 51, 56, 60, 62, 84}
$2^{2+11+22}.(M_{24} \times S_3)$	{32, 40, 48, 56, 60, 66, 69, 84, 88, 92}
$3^{1+12}.2Suz.2$	{36, 40, 45, 48, 54, 56, 60, 66, 78, 84}
$2^{5+10+20}.(S_3 \times L_5(2))$	{32, 40, 48, 56, 60, 62, 84, 93}
$S_3 \times Th$	{24, 36, 38, 54, 57, 60, 62, 78, 84, 93}
$2^{3+6+12+18}.(L_3(2) \times 3S_6)$	{32, 40, 48, 56, 60, 70, 84, 105}
$3^8.O_8^-(3).2_3$	{24, 27, 36, 39, 40, 41, 42, 45, 52, 56, 60}
$(D_{10} \times HN).2$	{24, 36, 38, 40, 44, 45, 50, 60, 70, 84, 95, 105, 110}
$(3^2:2 \times O_8^+(3)).S_4$	{24, 36, 40, 54, 56, 60, 78, 84, 104}
$3^{2+5+10}.(M_{11} \times 2S_4)$	{24, 36, 40, 45, 54, 60, 66, 88}
$3^{3+2+6+6}.(L_3(3) \times SD_{16})$	{24, 36, 54, 78, 104}
$5^{1+6}:2J_2:4$	{24, 40, 50, 56, 60, 70}
$(7:3 \times He):2$	{48, 51, 56, 60, 70, 84, 105, 119}
$(A_5 \times A_{12}):2$	{22, 24, 33, 36, 40, 45, 55, 60, 70, 84, 105}
$5^{3+3}.(2 \times L_3(5))$	{24, 50, 60, 62}
$(A_6 \times A_6 \times A_6).(2 \times S_4)$	{9, 24, 40, 60}
$(A_5 \times U_3(8):3_1):2$	{24, 28, 36, 38, 42, 45, 57, 60, 95, 105}
$5^{2+2+4}:(S_3 \times GL_2(5))$	{24, 50, 60}
$(L_3(2) \times S_4(4):2).2$	{40, 48, 51, 56, 60, 68, 70, 84, 105, 119}
$7^{1+4}:(3 \times 2S_7)$	{24, 56, 60, 70, 84}
$(5^2:[2^4] \times U_3(5)).S_3$	{24, 40, 60, 70, 84}
$(L_2(11) \times M_{12}).2$	{24, 40, 60, 66, 88, 110}
$(A_7 \times (A_5 \times A_5):2^2):2$	{24, 40, 60, 70, 84, 105}
$5^4:(3 \times 2L_2(25)):2_2$	{20, 24, 30, 78}
$7^{2+1+2}.GL_2(7)$	{42, 48, 56}
$M_{11} \times A_6.2^2$	{24, 30, 40, 66, 88, 110}
$(S_5 \times S_5 \times S_5):S_3$	{18, 24, 40, 60}
$(L_2(11) \times L_2(11)):4$	{24, 55, 60, 66}
$13^2:2L_2(13).4$	{12, 52, 56}
$(7^2:(3 \times 2A_4) \times L_2(7)).2$	{24, 84}
$(13:6 \times L_3(3)).2$	{24, 78, 104}
$13^{1+2}:(3 \times 4S_4)$	{24, 52, 78}
$L_2(71)$	{35, 36, 71}
$L_2(59)$	{29, 30, 59}
$11^2:(5 \times 2A_5)$	{20, 30, 55}

(continued)

Table 10.1 (continued)

Subgroup	Element orders
$L_2(29):2$	$\{28, 29, 30\}$
$7^2:SL_2(7)$	$\{6, 8, 14\}$
$L_2(19):2$	$\{18, 19, 20\}$
$41:40$	$\{40, 41\}$
$L_2(41)$	$\{20, 21, 41\}$
$L_2(13).2$ *	$\{12, 13, 14\}$
$L_2(16).4$ *	$\{8, 10, 12, 15, 17\}$

Using the information in Table 10.1, we now construct an auxiliary graph. We let Γ be the graph whose vertices are the orders of the elements of the Monster and where two vertices x and y are declared to be adjacent, if there exists no maximal subgroup M containing elements having order x and y. It can be verified, by magma, that $\{41, 59, 71, 87, 92, 93, 94, 95, 119\}$ is a clique of Γ and hence

$$\kappa(G) \geq 9. \tag{10.1}$$

Indeed, let $x_1, \ldots, x_9$ be the orders of elements forming a clique of size 9 in Γ and consider $g_i \in G$ such that $\mathbf{o}(g_i) = x_i$. The conjugacy classes $C_i = g_i^G$ are nine distinct vertices in $\Lambda(G)$. Pick $c_i \in C_i$ and $c_j \in C_j$ for $i \neq j$. Then $\langle c_i, c_j \rangle = G$, otherwise $\langle c_i, c_j \rangle$ would be contained in a maximal subgroup M of G. Since $\mathbf{o}(c_i) = x_i$ and $\mathbf{o}(c_j) = x_j$, by the definition of Γ, there is no edge incident to x_i, x_j in Γ, against the fact that they are part of a clique.

Next, we have computed the permutation characters of G acting on the cosets of its maximal subgroups $2.F_{2+}$, $2^{1+24}.Co_1$, $3.Fi_{24}$ and $2^{5+10+20}.(S_3 \times L_5(2))$. Recall that these are some of the permutation characters of primitive faithful actions of G already available in the literature. Using these characters, we have checked that, the elements of the Monster not belonging to some conjugate of these four maximal subgroups have order

$$41, 57, 59, 71, 95, 119.$$

Therefore, to obtain a normal covering of the Monster, using as components $2.F_{2+}$, $2^{1+24}.Co_1$, $3.Fi_{24}$ and $2^{5+10+20}.(S_3 \times L_5(2))$, we need to cover these remaining elements. It is important to observe that, for every $c \in \{41, 57, 59, 71, 95, 119\}$, G contains a unique conjugacy class[5] of cyclic subgroups of order c. Therefore,

---

[5] For $c \in \{41, 59, 71\}$, this conclusion follows from Sylow's theorem as c represents the largest prime power dividing $|G|$. In the case of $c = 59$, this is corroborated by [29] due to the unique conjugacy class of elements of order 57 within $G = F_1$. Similarly, for $c \in \{95, 119\}$, this can be inferred from [29]; specifically, G contains two conjugacy classes of elements of order c, and representatives for these classes can be selected as an element of order c along with its inverse.

consulting Table 10.1, we see that using

$$(A_5 \times U_3(8).3).2,\ (L_3(2) \times S_4(4)).2,\ L_2(71),\ L_2(59),\ 41:40,$$

we cover all the remaining elements.

Therefore, we obtain a normal covering of the Monster with 9 subgroups and hence

$$\gamma(G) \leq 9. \tag{10.2}$$

From (1.1), (10.1) and (10.2), we finally deduce $\gamma(G) = \kappa(G) = 9$.

For sporadic simple groups G admitting outer automorphisms, the computations for $\gamma_w(G)$, $\kappa_w(G)$, $\gamma(\mathrm{Aut}(G))$ and $\kappa(\mathrm{Aut}(G))$ are entirely similar. Indeed, we may compute $\gamma(\mathrm{Aut}(G))$ and $\kappa(\mathrm{Aut}(G))$ arguing exactly as above because the list of the maximal subgroups and the character table of $\mathrm{Aut}(G)$ is available and stored in GAP. For computing $\gamma_w(G)$, we determine the list of all the permutation characters of the actions of $\mathrm{Aut}(G)$ acting on the right cosets of M, where M is a maximal subgroup of G. Then, $\gamma_w(G)$ is simply the minimum number of such characters $\pi_1, \ldots, \pi_\ell$ such that the class function $\pi_1 + \cdots + \pi_\ell$ does not vanish in G. Similarly, using this list of permutation characters, we may also determine the edges of the Aut-invariably generating graph. Indeed, two conjugacy classes $x_1^{\mathrm{Aut}(G)}$ and $x_2^{\mathrm{Aut}(G)}$, with $x_1, x_2 \in G$, are adjacent in the Aut-invariably generating graph if and only if $\pi(x_1)\pi(x_2) = 0$, for every permutation character as above. Now, using built-in algorithms we may compute the clique number $\kappa_w(G)$ of this graph.

Chapter 11
Dropping the Maximality

In this chapter, we use Theorem 1.5 to classify the weak normal 2-coverings and the normal 2-coverings of the non-abelian simple groups. This means that we drop the hypothesis about the maximality of the components and give the description of all the possible components in a weak normal 2-covering of G, for G a non-abelian simple group such that $\gamma_w(G) = 2$.

Theorem 11.1 *Let G be a finite non-abelian simple group and let $\mu = \{H, K\}$ be a weak normal 2-covering of G. Then the pair (H, K) appears in Tables 11.1, 11.2, 11.3, 11.4, 11.5, 11.6, 11.7, 11.8, and 11.9, up to $\mathrm{Aut}(G)$-conjugacy. Moreover, μ gives rise to at least a normal 2-covering of G if and only if in the corresponding row of the fifth column of those tables appears a number greater than 0.*

From Theorem 11.1, we obtain the following rather remarkable corollary.

Corollary 11.2 *Let G be a finite non-abelian simple group and let $\mu = \{H, K\}$ be a weak normal 2-covering of G. Then either H or K is maximal in G, or $G \cong \mathrm{PSL}_3(4)$. When $G \cong \mathrm{PSL}_3(4)$, there exists a unique weak normal 2-covering $\{H, K\}$, with H and K not maximal in G, see Table 11.5.*

We adopt the same notation as in Notation 1.4 for reading Tables 11.1, 11.2, 11.3, 11.4, 11.5, 11.6, 11.7, 11.8, and 11.9. In these tables, when listing H and K, in the case that either H or K is not maximal in G, we also give some embeddings in certain overgroups. Observe that in Table 11.9, we are not listing all weak normal 2-coverings, because $\mathrm{P}\Omega_8^+(2)$ has 60 distinct $\mathrm{Aut}(\mathrm{P}\Omega_8^+(2))$-conjugacy classes of weak normal 2-coverings and $\mathrm{P}\Omega_8^+(3)$ has 2019 distinct $\mathrm{Aut}(\mathrm{P}\Omega_8^+(3))$-conjugacy classes of weak normal 2-coverings. It is quite remarkable that, although $\mathrm{P}\Omega_8^+(2)$ and $\mathrm{P}\Omega_8^+(3)$ have so many weak normal 2-coverings up to Aut-conjugacy, it can be checked with a computer that in each weak normal 2-covering $\{H, K\}$ either H or K is maximal. Actually more is true. Using the notation from [29], let x be an

Table 11.1 Components of the weak normal 2-coverings of non-abelian simple groups: alternating groups

Grp	Comp. H	Comp. K	Comments	N.	Nr.
A_5	$A_5 \cap (S_2 \times S_3)$	D_{10}		1	5
	A_4	D_{10}		1	
	$A_5 \cap (S_2 \times S_3)$	$5 < D_{10}$		1	
	A_4	$5 < D_{10}$		1	
	$A_3 < A_4$	D_{10}		1	
A_6	$A_6 \cap (S_2 \times S_4)$	A_5		2	8
	$A_6 \cap (S_3 \mathrm{wr} S_2)$	A_5		2	
	$D_8 < A_6 \cap (S_2 \times S_4)$	A_5	$|H|=8$	0	
	$A_6 \cap (S_3 \mathrm{wr} S_2)$	$D_{10} < A_5$		1	
	$A_6 \cap (S_2 \times S_4)$	$D_{10} < A_5$		0	
	$C_4 < D_8 < A_6 \cap (S_2 \times S_4)$	A_5	$|H|=4$	0	
	$A_6 \cap (S_3 \mathrm{wr} S_2)$	$5 < D_{10} < A_5$		1	
	$A_6 \cap (S_2 \times S_4)$	$5 < D_{10} < A_5$		0	
A_7	$A_7 \cap (S_2 \times S_5)$	$SL_3(2)$		2	2
	$A_7 \cap (S_2 \times S_5)$	$7:3 < SL_3(2)$		1	
A_8	$A_8 \cap (S_3 \times S_5)$	$2^3 : SL_3(2)$		2	4
	$A_3 \times A_5 < A_8 \cap (S_3 \times S_5)$	$2^3 : SL_3(2)$		2	
	$\langle (1\,2\,3)(4\,5\,6\,7\,8), (2\,3)(5\,7\,8\,6) \rangle <$ $A_8 \cap (S_3 \times S_5)$	$2^3 : SL_3(2)$	$|H|=60$	2	
	$A_3 \times D_{10} < A_8 \cap (S_3 \times S_5)$	$2^3 : SL_3(2)$	$|H|=30$	2	
A_9	$A_9 \cap (S_4 \times S_5)$	$P\Gamma L_2(8)$		0	1

Table 11.2 Components of the weak normal 2-coverings of non-abelian simple groups: sporadic groups, notation from [29]

Grp	Comp. H	Comp. K	Comments	N.	Nr.
M_{11}	$M_8 : S_3 \cong 2\dot{}S_4$	$PSL_2(11)$		1	8
	$M_9 : S_2 \cong 3^2 : Q_8.2$	$PSL_2(11)$		1	
	$M_{10} \cong A_6.2$	$PSL_2(11)$		1	
	$\langle (1\,5\,3\,8\,9\,2\,7\,11)(4\,10), (1\,6\,9)(2\,7\,8)$ $(3\,5\,11) \rangle < M_9 : S_2$	$PSL_2(11)$	$|H|=72$	1	
	$Q_8.2 < (M_8 : S_3) \cap (M_9 : S_2) \cap M_{10}$	$PSL_2(11)$	$|H|=16$	1	
	$C_8 < Q_8.2 < (M_8 : S_3) \cap (M_9 :$ $S_2) \cap M_{10}$	$PSL_2(11)$	$|H|=8$	1	
	$M_8 : S_3 \cong 2\dot{}S_4$	$C_{11} : C_5 < PSL_2(11)$		1	
	$M_{10} \cong A_6.2$	$C_{11} : C_5 < PSL_2(11)$		1	
M_{12}	$M_{10} : 2 \cong A_6.2^2$	$PSL_2(11)$		0	3
	M_{11}	$2 \times S_5$		0	
	M_{11}	$2 \times A_5 < 2 \times S_5$		0	

Table 11.3 Components of the weak normal 2-coverings of non-abelian simple groups: exceptional groups

Grp	Comp. H	Comp. K	Comments	N.	Nr.
$G_2(q)$	$SL_3(q).2$	$SU_3(q).2$	$q \geq 4, q$ even	1	1
$G_2(2)'$	$PSL_2(7)$	$4 \cdot S_4$	$G_2(2)' \cong PSU_3(3)$	1	4
	$PSL_2(7)$	$3_+^{1+2}:8$		1	
	$PSL_2(7)$	$3:8 < (4.S_4) \cap (3_+^{1+2}:8)$		1	
	$7:3 < PSL_2(7)$	$4.S_4$		1	
$G_2(3)$	$PSL_2(13)$	$[q^5]:GL_2(3)$		0	2
	$PSL_3(3):2$	$PSL_2(8):3$		0	
$^2G_2(3)'$	D_{18}	D_{14}	$^2G_2(3)' \cong PSL_2(8)$	1	5
	D_{18}	$2^3:7$		1	
	$C_9 < D_{18}$	$2^3:7$		1	
	$C_9 < D_{18}$	D_{14}		1	
	D_{18}	$C_7 < D_{14}$		1	
$^2F_4(2)'$	$2.[2^8]:5:4$	$PSL_3(3):2$		1	2
	$2.[2^8]:5:4$	$PSL_2(25)$		1	
$F_4(q)$	$^3D_4(q).3$	$Spin_9(q)$	$q = 3^a$	1	1

Table 11.4 Components of the weak normal 2-coverings of non-abelian simple groups: linear groups $PSL_2(q)$ (For $PSL_2(4) \cong A_5$, $PSL_2(5) \cong A_5$, $PSL_2(9) \cong A_6$, $PSL_4(2) \cong A_8$, see Table 11.1)

Grp	Comp. H	Comp. K	Comments	N.	Nr.
$PSL_2(7)$	S_4	Parabolic		2	4
	$D_8 < S_4$	Parabolic		1	
	$C_4 < D_8 < S_4$	Parabolic		1	
	S_4	$C_7 <$ Parabolic		2	
$PSL_2(q)$	D_{q+1}	Parabolic	q odd, $q > 9$	1	2
	$C_{(q+1)/2} < D_{q+1}$	Parabolic	q odd, $q > 9$	1	
$PSL_2(q)$	$D_{2(q+1)}$	Parabolic	$q > 4, q$ even	1	5
	$C_{q+1} < D_{2(q+1)}$	Parabolic	$q > 4, q$ even	1	
	$D_{2(q+1)}$	$C_{q-1} <$parabolic	$q > 4, q$ even	1	
	$D_{2(q+1)}$	$D_{2(q-1)}$	$q > 4, q$ even	1	
	$C_{q+1} < D_{2(q+1)}$	$D_{2(q-1)}$	$q > 4, q$ even	1	

element of order 9 in the class $9J$ of $P\Omega_8^+(3)$.[1] Then $\{\Omega_7(3), \langle x \rangle\}$ is a weak normal 2-covering of $P\Omega_8^+(3)$. In particular, for every subgroup H of $P\Omega_8^+(3)$ containing x, $\{\Omega_7(3), H\}$ is a weak normal 2-covering. It turns out that, up to $\mathrm{Aut}(P\Omega_8^+(3))$-conjugacy, there are 2011 weak normal 2-coverings of this type.

---

[1] Contrary to what one might expect at first, x is not a regular unipotent element. Indeed, x has Jordan blocks of size 3 and 5.

Table 11.5 Components of the weak normal 2-coverings of non-abelian simple groups: linear groups $\mathrm{PSL}_3(q)$ (For $\mathrm{PSL}_3(2)$, use $\mathrm{PSL}_3(2) \cong \mathrm{PSL}_2(7)$)

Grp	Comp. H	Comp. K	Comments	N.	Nr.
$\mathrm{PSL}_3(q)$	$\left(\frac{q^2+q+1}{\gcd(3,q-1)}\right):3$	$E_q^2 : \mathrm{GL}_2(q)$ parabolic	$\gcd(3,q)=1, q>4$	2	2
	$\frac{q^2+q+1}{\gcd(3,q-1)}$	$E_q^2 : \mathrm{GL}_2(q)$ parabolic	$\gcd(3,q)=1, q>4$	2	
$\mathrm{PSL}_3(3)$	$13:3$	$E_3^2 : \mathrm{GL}_2(3)$ parabolic		2	4
	13	$E_3^2 : \mathrm{GL}_2(3)$ parabolic		2	
	$13:3$	$E_3^2 : (8:2) < E_3^2 : \mathrm{GL}_2(3)$		2	
	$13:3$	$\mathrm{GL}_2(3) < E_3^2 : \mathrm{GL}_2(3)$		1	
$\mathrm{PSL}_3(q)$	$\left(\frac{q^2+q+1}{\gcd(3,q-1)}\right):3$	$E_q^2 : \mathrm{GL}_2(q)$ parabolic	$\gcd(3,q)=3, q>3$	2	3
	$\frac{q^2+q+1}{\gcd(3,q-1)}$	$E_q^2 : \mathrm{GL}_2(q)$ parabolic	$\gcd(3,q)=3, q>3$	2	
	$\frac{q^2+q+1}{\gcd(3,q-1)}:3$	$\mathrm{GL}_2(q) < E_q^2 : \mathrm{GL}_2(q)$	$\gcd(3,q)=3, q>3$	1	
$\mathrm{PSL}_3(4)$	$\mathrm{SL}_3(2)$	A_6		0	13
	$\mathrm{SL}_3(2)$	Parabolic		6	
	$7:3 < \mathrm{SL}_3(2)$	Parabolic		2	
	$\mathrm{SL}_3(2)$	$K <$ parabolic	$|K|=60$	0	
	$\mathrm{SL}_3(2)$	$K <$ parabolic	$|K|=60$	0	
	$\mathrm{SL}_3(2)$	$K <$ parabolic	$|K|=80$	0	
	$\mathrm{SL}_3(2)$	$K <$ parabolic	$|K|=160$	6	
	$7:3 < \mathrm{SL}_3(2)$	$K <$ parabolic	$|K|=160$	2	
	$\mathrm{SL}_3(2)$	$5:2$		0	
	$\mathrm{SL}_3(2)$	$5 < 5:2$		0	
	$7:3 < \mathrm{SL}_3(2)$	A_6		0	
	$7 < 7:3 < \mathrm{SL}_3(2)$	A_6		0	
	$7 < 7:3 < \mathrm{SL}_3(2)$	Parabolic		2	

Table 11.6 Components of the weak normal 2-coverings of non-abelian simple groups: linear groups $\mathrm{PSL}_4(q)$ (For $\mathrm{PSL}_4(2)$, use $\mathrm{PSL}_4(2) \cong A_8$)

Grp	Comp. H	Comp. K	Comments	N.	Nr.
$\mathrm{PSL}_4(q)$	$\frac{1}{d}\mathrm{SL}_2(q^2).(q+1).2$	$\frac{1}{d}E_q^3 : \mathrm{GL}_3(q)$	$d := \gcd(4,q-1), q \geq 4$	2	2
	$\frac{1}{d}\mathrm{SL}_2(q^2).(q+1) <$ $\frac{1}{d}\mathrm{SL}_2(q^2).(q+1).2$	$\frac{1}{d}E_q^3 : \mathrm{GL}_3(q)$	$d := \gcd(4,q-1), q \geq 4$	2	
$\mathrm{PSL}_4(3)$	$\frac{1}{2}\mathrm{SL}_2(9).4.2$	$\frac{1}{2}E_3^3 : \mathrm{GL}_3(3)$		2	4
	$\frac{1}{2}\mathrm{SL}_2(9).4 < \frac{1}{2}\mathrm{SL}_2(9).4.2$	$\frac{1}{2}E_3^3 : \mathrm{GL}_3(3)$		2	
	$(2 \times A_5).2^2 < \frac{1}{2}\mathrm{SL}_2(9).4.2$	$\frac{1}{2}E_3^3 : \mathrm{GL}_3(3)$		0	
	$A_5.4 < (2 \times A_5).2^2 <$ $\frac{1}{2}\mathrm{SL}_2(9).4.2$	$\frac{1}{2}E_3^3 : \mathrm{GL}_3(3)$		0	

11 Dropping the Maximality

Table 11.7 Components of the weak normal 2-coverings of non-abelian simple groups: unitary groups

Grp	Comp. H	Comp. K	Comments	N.	Nr.
$PSU_3(q)$	$(q^2 - q + 1) : 3$	$GU_2(q)$	$3 = \gcd(q, 3), q > 3$	1	1
$PSU_3(3)$	$PSL_2(7)$	$GU_2(3)$		1	4
	$PSL_2(7)$	$E_3^{1+2} : 8$		1	
	$PSL_2(7)$	$3 : 8 < GU_2(3) \cap E_3^{1+2} : 8$		1	
	$7 : 3 < PSL_2(7)$	$GU_2(3)$		1	
$PSU_3(5)$	A_7	$\frac{1}{3}GU_2(5)$		0	3
	A_7	$\frac{1}{3}E_5^{1+2} : 24$		3	
	A_7	$4 : 8 < \frac{1}{3}E_5^{1+2} : 24$	$|K| = 40$	0	
$PSU_4(q)$	$\frac{1}{d}GU_3(q)$	$\frac{1}{d}E_q^4 : SL_2(q^2) : (q-1)$	$d := \gcd(4, q+1), q \geq 4$	1	1
$PSU_4(2)$	$GU_3(2)$	$Sp_4(2)$		1	3
	$GU_3(2)$	$E_2^4 : SL_2(4)$		1	
	$GU_3(2)$	$S_5 < Sp_4(2)$	$|K| = 120$	1	
$PSU_4(3)$	A_7	$\frac{1}{4}E_3^{1+4} : SU_2(3) : 8$		4	4
	$\frac{1}{4}GU_3(3)$	$\frac{1}{4}E_3^4 : SL_2(9) : 2$		1	
	$PSL_3(4)$	$\frac{1}{4}E_3^{1+4} : SU_2(3) : 8$		2	
	$\frac{1}{4}GU_3(3)$	$PSU_4(2)$		0	
$PSU_6(2)$	$Sp_6(2)$	$PGU_5(2)$		0	2
	$PSU_4(3) : 2$	$PGU_5(2)$		0	

Table 11.8 Components of the weak normal 2-coverings of non-abelian simple groups: symplectic groups (For $PSp_4(2)' \cong A_6$, see Table 1.3)

Group	Comp. H	Comp. K	Comments	N.	Nr.
$PSp_4(3)$	$\frac{1}{2}E_3^{1+2} : (2 \times Sp_2(3))$	$PSp_2(9) : 2$		1	3
	$\frac{1}{2}E_3^{1+2} : (2 \times Sp_2(3))$	$2^4.A_5$		1	
	$\frac{1}{2}E_3^{1+2} : (2 \times Sp_2(3))$	$S_5 < PSp_2(9) : 2$		1	3
$Sp_n(q)$	$SO_n^-(q)$	$SO_n^+(q)$	$n \geq 6, q$ even	1	1
$Sp_4(q)$	$SO_4^-(q)$	$SO_4^+(q)$	$q \geq 8, q$ even	2	2
	$(q^2+1).[4] < SO_4^-(q)$	$SO_4^+(q)$		0	
$Sp_4(4)$	$SO_4^-(4)$	$SO_4^+(4)$		2	8
	$Sp_2(16) : 2$	$Sp_4(2)$		0	
	$Sp_2(16) : 2$	$Sp_4(2)' < Sp_4(2)$		0	
	$SO_4^-(4)$	$K < SO_4^+(4)$	$|K| = 200$	0	
	$SO_4^-(4)$	$K < SO_4^+(4)$	$|K| = 100$	0	
	$Sp_2(16) : 2$	$S_5 < SO_4^+(4) \cap Sp_4(2)$		0	
	$Sp_2(16) : 2$	$5 : 4 < S_5 < SO_4^+(4) \cap Sp_4(2)$		0	
	$7 : 4 < SO_4^-(4)$	$SO_4^+(4)$		0	
$PSp_6(3^f)$	$\frac{1}{2}(Sp_2(3^f) \perp Sp_4(3^f))$	$\frac{1}{2}Sp_2(3^{3f}) : 3$		1	1

Table 11.9 Components of the weak normal 2-coverings of non-abelian simple groups: orthogonal groups

Group	Comp. H	Comp. K	Comments	N.	Nr.
$P\Omega_8^+(2)$	–	–	Omitted	0	60
$P\Omega_8^+(3)$	–	–	Omitted	0	2019

We now give the proof of Theorem 11.1. Let G be a non-abelian simple group and let $\mu = \{H, K\}$ be a weak normal 2-covering of G. Let H_M and K_M be maximal subgroups of G with $H \leq H_M$ and $K \leq K_M$. Then $\{H_M, K_M\}$ is a weak normal 2-covering of G with maximal components and, by Theorem 1.5, the group G and the pair (H_M, K_M) appear in Tables 1.3–1.7.

Except when (G, H_M, K_M) is part of an infinite family of weak normal 2-coverings, we may simply use the computer algebra system magma to prove Theorem 11.1. Hence, for proving Theorem 11.1, it suffices to consider the following cases:

(1) $G = G_2(q)$, $H_M \cong \mathrm{SL}_3(q).2$, $K_M \cong \mathrm{SU}_3(q).2$, $q \geq 4$ and q even;
(2) $G = F_4(q)$, $H_M \cong {}^3D_4(q).3$, $K_M \cong \mathrm{Spin}_9(q) \cong 2.\Omega_9(q)$, $q = 3^a$, $a \geq 1$;
(3) $G = \mathrm{PSL}_2(q)$, $H_M \cong D_{q+1}$, $K_M \cong E_q : ((q-1)/2)$, $q > 9$ and q odd;
(4) $G = \mathrm{PSL}_2(q)$, $H_M \cong D_{2(q+1)}$, $K_M \cong E_q : (q-1)$, $q > 4$ and q even;
(5) $G = \mathrm{PSL}_2(q)$, $H_M \cong D_{2(q+1)}$, $K_M \cong D_{2(q-1)}$, $q > 4$ and q even;
(6) $G = \mathrm{PSL}_3(q)$, $H_M \cong \frac{q^2+q+1}{\gcd(3,q-1)} : 3$, K_M parabolic and $q \neq 4$;
(7) $G = \mathrm{PSL}_4(q)$, $H_M \cong \frac{1}{d}\mathrm{SL}_2(q^2).(q+1).2$, $K_M \cong \frac{1}{d}E_q^3 : \mathrm{GL}_3(q)$ and $d = \gcd(4, q-1)$;
(8) $G = \mathrm{PSU}_3(q)$, $H_M \cong (q^2 - q + 1) : 3$, $K_M \cong \mathrm{GU}_2(q)$, $q = 3^a$ and $a > 1$;
(9) $G = \mathrm{PSU}_4(q)$, $H_M \cong \frac{1}{d}\mathrm{GU}_3(q)$, $K_M \cong \frac{1}{d}E_q^4 : \mathrm{SL}_2(q^2) : (q-1)$, $d = \gcd(4, q+1)$ and $q \geq 4$;
(10) $G = \mathrm{Sp}_n(q)$, $H_M = \mathrm{SO}_n^-(q)$, $K_M = \mathrm{SO}_n^+(q)$, $n \geq 4$ even, q even and $(n, q) \neq (4, 2)$;
(11) $G = \mathrm{PSp}_6(q)$, $H_M = \frac{1}{2}(\mathrm{Sp}_2(q) \perp \mathrm{Sp}_4(q))$, $K_M = \frac{1}{2}\mathrm{Sp}_2(q^3) : 3$ and $q = 3^a$.

If $H = H_M$ and $K = K_M$, then the proof follows from Theorem 1.5. Therefore we may suppose that either

$$H < H_M \quad \text{or} \quad K < K_M. \tag{11.1}$$

Recall that for $X \leq G$, we use the notation $\tilde{X}$ to denote $\pi^{-1}(X)$. By Sect. 2.2, we have the weak normal 2-covering of $\tilde{G}$ given by $\tilde{\mu} = \{\tilde{H}, \tilde{K}\}$. Moreover, $\tilde{H}_M := \pi^{-1}(H_M)$, $\tilde{K}_M := \pi^{-1}(K_M)$ are maximal subgroups of $\tilde{G}$ containing $\tilde{H}$ and $\tilde{K}$ respectively and $\tilde{\mu}_M := \{\tilde{H}_M, \tilde{K}_M\}$ is a weak normal 2-covering of $\tilde{G}$ with maximal components.

11.1 Linear Groups: Cases (3)–(7)

Suppose $G = \mathrm{PSL}_2(q)$ and We Are in Case (3) or (4) Set $d := \gcd(2, q-1)$. Here, we have

$$H \leq H_M \cong D_{2(q+1)/d} \text{ and } K \leq K_M \cong E_q : ((q-1)/d).$$

Now, G contains an element x of order $(q+1)/d$ and an element y of order $(q-1)/d$. As K_M contains no element having order $(q+1)/d$ and H_M contains no element having order $(q-1)/d$, up to replacing x and y with suitable $\mathrm{Aut}(G)$-conjugates, we deduce

$$C_{(q+1)/d} \cong \langle x \rangle \leq H \leq H_M = \mathbf{N}_G(\langle x \rangle) \cong D_{2(q+1)/d},$$
$$C_{(q-1)/d} \cong \langle y \rangle \leq K \leq K_M \cong E_q : ((q-1)/d).$$

Observe that $\langle x \rangle$ is a maximal subgroup of H_M and $\langle y \rangle$ is a maximal subgroup of K_M. It follows that $H \in \{\langle x \rangle, H_M\}$ and $K \in \{\langle y \rangle, K_M\}$.

Assume now q odd. Then, H contains no non-identity unipotent elements. Therefore, K contains a non-identity unipotent element u. Thus $\langle y \rangle < \langle y, u \rangle \leq K$. From the maximality of $\langle y \rangle$ in K_M, we get $K_M = \langle y, u \rangle = K$. Now, (11.1) gives $H = \langle x \rangle$. It is easy to verify that $\langle x \rangle$ and $K_M \cong E_q : ((q-1)/2)$ is a normal 2-covering of G.

Assume now q even. If H contains an involution u, then $H = \langle x, u \rangle = \mathbf{N}_G(\langle x \rangle) = H_M \cong D_{2(q+1)}$ so that, again by (11.1), we get $K = \langle y \rangle$ and it is readily seen that $\{H_M, \langle y \rangle\}$ is a normal 2-covering of G. If H contains no involution, then $H = \langle x \rangle \cong C_{q+1}$, K must contain some involutions u and $K = \langle y, u \rangle = K_M \cong E_q : (q-1)$. Again, it is readily seen that $\{\langle x \rangle, K_M\}$ is a normal 2-covering of G.

Suppose $G = \mathrm{PSL}_2(q)$ and We Are in Case (5) Here, q is even and we have $H \leq H_M \cong D_{2(q+1)}$ and $K \leq K_M \cong D_{2(q-1)}$. As above, G contains an element x of order $q+1$ and an element y of order $q-1$ and we have

$$\langle x \rangle \leq H \leq H_M = \mathbf{N}_G(\langle x \rangle) \cong D_{2(q+1)},$$
$$\langle y \rangle \leq K \leq K_M = \mathbf{N}_G(\langle y \rangle) \cong D_{2(q-1)}.$$

Since either H or K contains elements of order 2, we have $H = H_M$ or $K = K_M$. If $K < K_M$, we obtain $H = H_M \cong D_{2(q+1)}$ and $K = \langle y \rangle \cong C_{q-1}$. Note that the covering $\{H_M, \langle y \rangle\}$ has already been considered in the paragraph above. If $K = K_M$, we instead obtain one additional normal 2-covering with $H = \langle x \rangle \cong C_{q+1}$.

Suppose $G = \mathrm{PSL}_3(q)$ and We Are in Case (6) Here it is more convenient to work with $\tilde{G} = \mathrm{SL}_3(q)$ and the covering $\tilde{\mu} = \{\tilde{H}, \tilde{K}\}$. We have $\tilde{H} \leq \tilde{H}_M \cong (q^2 + q + 1) : 3$ and $\tilde{K} \leq \tilde{K}_M \cong E_q^2 : \mathrm{GL}_2(q)$ is parabolic. Now, $\tilde{G}$ contains an element x of order $q^2 + q + 1$ and an element y of order $q^2 - 1$. As $\tilde{K}_M$ contains no element having order $q^2 + q + 1$ and $\tilde{H}_M$ contains no element having order $q^2 - 1$, up to replacing x and y with suitable $\mathrm{Aut}(\tilde{G})$-conjugates, we deduce

$$C_{q^2+q+1} \cong \langle x \rangle \leq \tilde{H} \leq \tilde{H}_M = \mathbf{N}_{\tilde{G}}(\langle x \rangle) \cong (q^2 + q + 1) : 3,$$

$$C_{q^2-1} \cong \langle y \rangle \leq \tilde{K} \leq \tilde{K}_M \cong E_q^2 : \mathrm{GL}_2(q).$$

Observe that C_{q^2+q+1} is a maximal subgroup of $\tilde{H}_M$. Thus $\tilde{H} \in \{\langle x \rangle, \mathbf{N}_{\tilde{G}}(\langle x \rangle)\}$.

Assume first that q is not a power of 3, or that $\tilde{H} = \langle x \rangle \cong C_{q^2+q+1}$. Observe that $\tilde{H}$ has no non-identity unipotent elements. Therefore, all non-identity unipotent elements of $\tilde{G}$ have an $\mathrm{Aut}(\tilde{G})$-conjugate in $\tilde{K}$. Let $\tilde{P} \cong E_q^2$ be the Fitting subgroup of $\tilde{K}_M$. Note that the non-identity unipotent elements in $\tilde{P}$ are not regular. Since $\tilde{G}$ contains regular unipotent elements, $\tilde{K}$ contains a regular unipotent element u and $u \in \tilde{K}_M \setminus \tilde{P}$. Now, using the subgroup structure of $\mathrm{GL}_2(q)$ and using the fact that $\langle u, y \rangle \leq \tilde{K}$, we deduce that $\tilde{K}_M = \langle y, u, \tilde{P} \rangle$, that is, $\tilde{K}\tilde{P} = \tilde{K}_M$. Since $\mathrm{GL}_2(q)$ acts irreducibly on E_q^2, we deduce that $\mathrm{GL}_2(q)$ is a maximal subgroup of $E_q^2 : \mathrm{GL}_2(q)$. Therefore, either $\tilde{K} = \tilde{K}_M \cong E_q^2 : \mathrm{GL}_2(q)$ or $\tilde{K} \cong \mathrm{GL}_2(q)$. In the former case, routine considerations of the possible Jordan forms of the elements in $\tilde{G} = \mathrm{SL}_3(q)$ show that $\{\langle x \rangle, \tilde{K} = \tilde{K}_M\}$ is a normal 2-covering of $\tilde{G}$. In the latter case, $\tilde{K}$ is a complement of $\tilde{P}$ in $\tilde{K}_M$. Since all complements of $\tilde{P}$ in $\tilde{K}_M$ are $\tilde{K}_M$-conjugate, see [28, Theorem (4.2) and Table (4.5)],[2] we may suppose that

$$\tilde{K} = \left\{ \begin{pmatrix} \det(A)^{-1} & 0 \\ 0 & A \end{pmatrix} \mid A \in \mathrm{GL}_2(q) \right\}.$$

However, we see that $\tilde{K}$ contains no regular unipotent elements, which is a contradiction. Therefore no 2-covering arises in this case.

Assume now that q is a power of 3 and that $\tilde{H} = \mathbf{N}_{\tilde{G}}(\langle x \rangle) \cong C_{q^2+q+1} : C_3$. When $q = 3$, the proof follows with a computer computation; therefore for the rest of the argument we suppose $q > 3$. This will help for avoiding some degenerate cases. Here the argument is rather similar to the argument above, only slightly more delicate. By (11.1), we have $\tilde{K} < \tilde{K}_M$. It can be shown that the non-identity unipotent elements of $\tilde{H}$ are regular.[3] In particular, all non-identity non-regular unipotent elements of $\tilde{G}$ have an $\mathrm{Aut}(\tilde{G})$-conjugate in $\tilde{K}$. Thus, $\tilde{K}$ contains a non-

---

[2] Strictly speaking, the work in [28] deals with the Lie group $\mathrm{SL}_n(q)$, however, the results there can be used also for deducing the corresponding result for $\mathrm{GL}_n(q)$.

[3] To see this, we may identify the $\mathbb{F}_q$-vector space $V = \mathbb{F}_{q^3}$ with the additive group of the field $\mathbb{F}_{q^3}$. Under this identification, the element $x \in \tilde{H}$ acts as a non-zero scalar of $\mathbb{F}_{q^3}$, and C_3 acts via field automorphisms. Let $1 \neq u \in \tilde{H}$ be a unipotent element, that is, u has order 3. Clearly, $\gcd(q^2 + q + 1, 3) = 1$. By replacing u with a suitable $\tilde{H}$-conjugate, we may suppose that u itself acts on $V = \mathbb{F}_{q^3}$ as a Galois automorphism of order 3. Clearly, $\mathbf{C}_V(u) = \mathbf{C}_{\mathbb{F}_{q^3}}(u) = \mathbb{F}_q$. This shows that $\dim_{\mathbb{F}_q} \mathbf{C}_V(u) = 1$, and hence, since $\dim_{\mathbb{F}_q}(V) = 3$, we deduce that u is a regular unipotent.

11.1 Linear Groups

identity non-regular unipotent element. As above, let $\tilde{P}$ be the Fitting subgroup of $\tilde{K}_M$. Observe that $\langle y \rangle \cong C_{q^2-1}$ acts by conjugation transitively on the non-identity elements of $\tilde{P}$. As a consequence, if there exists a non-identity non-regular unipotent element $u \in \tilde{P} \cap \tilde{K}$, then $\tilde{P} \leq \tilde{K}$. Using this observation we now locate the non-identity non-regular unipotent elements in $\tilde{K}$ with respect to $\tilde{P}$.

We first suppose that there exists a non-identity element $u \in \tilde{P} \cap \tilde{K}$. Then $\tilde{K} \geq \langle \tilde{P}, y \rangle \cong E_q^2 : (q^2 - 1)$. All the elements in $E_q^2 : (q^2 - 1)$ having order $q - 1$ are Aut$(\tilde{G})$-conjugate to an element in $\langle y^{q+1} \rangle$ and hence they have two distinct eigenvalues b and b^{-2} in $\mathbb{F}_q$ and their matrices are conjugate to

$$\begin{pmatrix} b^{-2} & 0 & 0 \\ 0 & b & 0 \\ 0 & 0 & b \end{pmatrix},$$

with $b \in \mathbb{F}_q^*$ having order $q - 1$. Since $\tilde{G}$ contains elements having order $q - 1$ and having three distinct eigenvalues in $\mathbb{F}_q$ (here we are using $q > 3$) and since $\tilde{H}$ has no elements of order $q - 1$, we deduce that $\tilde{K}$ contains a representative of each Aut$(\tilde{G})$-conjugacy class of elements having order $q - 1$ and having three distinct eigenvalues in $\mathbb{F}_q$. Let λ be a generator of the multiplicative group $\mathbb{F}_q^*$ and consider

$$z' := \begin{pmatrix} 1 & 0 & 0 \\ 0 & \lambda & 0 \\ 0 & 0 & \lambda^{-1} \end{pmatrix}.$$

Since q is a power of 3 with $q > 3$, it is not hard to verify that

$$\{1, \lambda, \lambda^{-1}\} \cap \{-1, -\lambda, -\lambda^{-1}\} = \emptyset. \tag{†}$$

Let $z \in \tilde{K}$ with z Aut$(\tilde{G})$-conjugate to z'. Observe that z has three distinct eigenvalues $1, \lambda, \lambda^{-1}$. We now use the subgroup structure of $\mathrm{GL}_2(q) \cong \tilde{K}_M/\tilde{P}$ to prove that $\tilde{K}$ cannot contain such an element. There exists only one proper subgroup of $\mathrm{GL}_2(q)$ properly containing a cyclic subgroup of order $q^2 - 1$, that is, the normalizer of this subgroup of order $q^2 - 1$. In other words, as $\tilde{P} \leq \tilde{K}$, we have $\tilde{K} \geq \langle y, u, z \rangle \cong E_q^2 : ((q^2 - 1).2)$. We identify $(q^2 - 1).2$ with the group $\mathbb{F}_{q^2}^* \rtimes \langle \phi \rangle$, where $\phi : \mathbb{F}_{q^2} \to \mathbb{F}_{q^2}$ is the Frobenius map $x \mapsto x^q$. The elements in $\mathbb{F}_{q^2}^*$ having order $q - 1$ lie in $\mathbb{F}_q^*$ and hence have a unique eigenvalue with multiplicity 2. Therefore our putative element z, interpreted as an element of $\mathbb{F}_{q^2}^* \rtimes \langle \phi \rangle$, can be written as $\lambda^\ell \phi$, for some $\ell \in \{1, \ldots, q^2 - 1\}$. Let μ_1 and μ_2 be the two distinct eigenvalues of $\lambda^\ell \phi$. Now,

$$\lambda^\ell \phi \lambda^\ell \phi = \lambda^\ell \lambda^{\ell q} = \lambda^{\ell(q+1)} \in \mathbb{F}_q.$$

Therefore, $\lambda^\ell \phi$ squares to an element in $\mathbb{F}_q^*$ and hence $\mu_1^2 = \mu_2^2$. This implies $\mu_2 \in \{\mu_1, -\mu_1\}$, contradicting (†).

The only remaining case is that all the non-identity non-regular unipotent elements in $\tilde{K}$ lie in $\tilde{K} \setminus \tilde{P}$. Thus such elements must be $\text{Aut}(\tilde{G})$-conjugate to an element of $\tilde{H}$. It follows that $3 \mid |\tilde{H}|$ so that $\tilde{H} = \tilde{H}_M \cong (q^2 + q + 1) : 3$ and, by (11.1), $\tilde{K} < \tilde{K}_M$. As we have seen above (when dealing with the case q not a power of 3), this yields $\tilde{K} \cong \text{GL}_2(q)$. Therefore, $\tilde{K}$ is a complement of $\tilde{P}$ in $\tilde{K}_M$. Since all complements of $\tilde{P}$ in $\tilde{K}$ are conjugate, see [28, Theorem (4.2) and Table (4.5)], we may suppose that

$$\tilde{K} = \left\{ \begin{pmatrix} \det(A)^{-1} & 0 \\ 0 & A \end{pmatrix} \mid A \in \text{GL}_2(q) \right\}.$$

Routine considerations on the possible Jordan forms of the elements of $\tilde{G} = \text{SL}_3(q)$ yield that $H = \mathbf{N}_{\tilde{G}}(\langle x \rangle) \cong (q^2+q+1) : 3$ and $\tilde{K} \cong \text{GL}_2(q)$ is a normal 2-covering.

Suppose $G = \text{PSL}_4(q)$ and We Are in Case (7) The case $q = 2$ is of no concern because $\text{PSL}_4(2) \cong A_8$. The weak normal coverings of $\text{PSL}_4(3)$ have been found with the auxiliary help of a computer. Therefore, for the rest of this case, we suppose $q \geq 4$. Here it is more convenient to work with $\tilde{G} = \text{SL}_4(q)$. We have $\tilde{H} \leq \tilde{H}_M \cong \text{SL}_2(q^2).(q+1).2$ and $\tilde{K} \leq \tilde{K}_M \cong E_q^3 : \text{GL}_3(q)$ is parabolic. The argument is very similar to the argument dealing with $\text{PSL}_3(q)$, but with some minor differences. Now, $\tilde{G}$ contains an element x of order $q^3 + q^2 + q + 1$ and an element y of order $q^3 - 1$. As $\tilde{K}_M$ contains no element having order $q^3 + q^2 + q + 1$ and $\tilde{H}_M$ contains no element having order $q^3 - 1$, up to replacing x and y with suitable $\text{Aut}(\tilde{G})$-conjugates, we deduce

$$C_{q^3+q^2+q+1} \cong \langle x \rangle \leq \tilde{H} \leq \tilde{H}_M \cong \text{SL}_2(q^2).(q+1).2,$$
$$C_{q^3-1} \cong \langle y \rangle \leq \tilde{K} \leq \tilde{K}_M \cong E_q^3 : \text{GL}_3(q).$$

Assume first that q is odd. We claim that $\tilde{K} = \tilde{K}_M$. Since $2(q + 1)$ is relatively prime to q, the non-identity unipotent elements of $\tilde{H}$ all lie in $\text{SL}_2(q^2)$. The Jordan form of the non-identity unipotent elements of $\text{SL}_2(q^2)$, viewed as 2×2-matrices over $\mathbb{F}_{q^2}$, is

$$\begin{pmatrix} 1 & 1 \\ 0 & 1 \end{pmatrix}.$$

These matrices, viewed as 4×4-matrices over $\mathbb{F}_q$, have Jordan form

$$\begin{pmatrix} 1 & 1 & 0 & 0 \\ 0 & 1 & 0 & 0 \\ 0 & 0 & 1 & 1 \\ 0 & 0 & 0 & 1 \end{pmatrix}.$$

11.1 Linear Groups

In particular, $\tilde{H}$ has no regular unipotent elements. Therefore $\tilde{K}$ contains a regular unipotent element u. Let $\tilde{P} \cong E_q^3$ be the Fitting subgroup of $\tilde{K}_M$. Using the subgroup structure of $\mathrm{GL}_3(q)$, which can be deduced from the subgroup structure of $\mathrm{SL}_3(q)$ in [9, Tables 8.3 and 8.4, p. 378], it is not hard to verify that one of the following holds:

- $\langle \tilde{P}, y, u \rangle = \tilde{K}_M$,
- q is a power of 3 and $\langle \tilde{P}, y, u \rangle / \tilde{P} \cong (q^3 - 1) : 3$.

In the first case, as we have seen similarly a few times in the case of $\tilde{G} = \mathrm{SL}_3(q)$, we deduce that either $\tilde{K} = \tilde{K}_M$ or $\tilde{K} \cong \mathrm{GL}_3(q)$ is a complement of $\tilde{P}$ in $\tilde{K}$. In the first possibility our claim is proved. In the second possibility, using [28, Theorem (4.2) and Table (4.5)], we have that all the complements of $\tilde{P}$ in $\tilde{K}_M$ are conjugate and hence $\tilde{K}$ is conjugate to

$$\left\{ \begin{pmatrix} \det(A)^{-1} & 0 \\ 0 & A \end{pmatrix} \mid A \in \mathrm{GL}_3(q) \right\}.$$

Since $\mathrm{GL}_3(q) \leq \tilde{G} = \mathrm{SL}_4(q)$ has no element having Jordan form consisting only of one block, we deduce that u cannot lie in a conjugate to $\mathrm{GL}_3(q)$ and hence we reach a contradiction.

In the second case, that is when q is a power of 3 and $\langle \tilde{P}, y, u \rangle / \tilde{P} \cong (q^3 - 1) : 3$, the same argument gives that either $\tilde{K} = \tilde{K}_M$ and our claim is proved, or $\tilde{K} = \langle y, u \rangle \cong E_q^3 : (q^3 - 1) : 3$. If $\tilde{K} = \langle y, u \rangle$, we prove that we reach a contradiction. Recall that $q > 3$. Let $z \in \tilde{G} = \mathrm{SL}_4(q)$, having type $2 \oplus 1 \oplus 1$, with $\mathbb{F}_q^4 = V_1 \oplus V_2 \oplus V_3$ with z inducing a matrix s of order $q^2 - 1$ on V_1, the 1×1 matrix $(\det(s))^{-1}$ on V_2 and the 1×1 matrix 1 on V_3. We show that any $\mathrm{Aut}(\tilde{G})$-conjugate of z is neither in $\tilde{H}_M$ nor in $\tilde{K}_M$. For $\tilde{K}$ this comes immediately from the fact that $\mathbf{o}(z) = q^2 - 1 \nmid |\tilde{K}|$. Assume next, by contradiction, that $z \in \tilde{H}_M = (\mathrm{GL}_2(q^2).2) \cap \mathrm{SL}_4(q)$. Then $z^2 \in \mathrm{GL}_2(q^2)$. Observe now that s^2 acts irreducibly on V_1. Indeed, by [14, Lemma 2.4], the minimal polynomial of s^2 is irreducible of degree $r \mid 2$, where $r \in \mathbb{N}$ is maximum such that $\frac{q^2-1}{q^r-1} \mid 2$. The possibility $r = 1$ implies the contradiction $q + 1 \leq 2$, so that $r = 2$. Moreover

$$\mathbf{o}((\det(s))^{-2}) = \frac{q-1}{2} \neq 1,$$

because $q \neq 3$. As a consequence, the action of z^2 is of type $2 \oplus 1 \oplus 1$ and 1 is an eigenvalue of z^2 of multiplicity one. Since no matrix in $\mathrm{GL}_2(q^2)$ can have 1 as eigenvalue of multiplicity one, we have reached a contradiction.

Therefore, for every q odd, we have proved our claim, that is, $\tilde{K} = \tilde{K}_M$.

Again, under the assumption that q is odd, we claim that $\mathrm{SL}_2(q^2).(q+1) = \tilde{H}$; recall $q > 3$. From (11.1), it suffices to show that $\mathrm{SL}_2(q^2).(q+1) \leq \tilde{H}$. Let w be an element of order $q^2 - 1$ in $\tilde{G} = \mathrm{SL}_4(q)$, having type $2 \oplus 2$ and such that $\mathbb{F}_q^4$ has only two distinct irreducible $\mathbb{F}_q\langle z \rangle$-subspaces. To construct such an element w

it suffices to take a Singer cycle s of $\mathrm{GL}_2(q)$ and consider $w = s \oplus (s^{-1})^T$. Now, by order considerations, w has no $\mathrm{Aut}(\tilde{G})$-conjugate in $\tilde{K}$ and hence w has an $\mathrm{Aut}(\tilde{G})$-conjugate in $\tilde{H}$. Without loss of generality, we may suppose that $w \in \tilde{H}$. Using the subgroup structure of $\mathrm{SL}_2(q^2)$ and recalling that $q^2 \neq 9$, it is not hard to verify that $\langle x^{q+1}, w \rangle \cong \mathrm{SL}_2(q^2)$. Therefore, $\tilde{H} \geq \langle x, w \rangle \geq \mathrm{SL}_2(q^2).(q+1)$.

Routine computations with the possible Jordan forms of the elements of $\mathrm{SL}_4(q)$ show that $\tilde{H} \cong \mathrm{SL}_2(q^2).(q+1)$ and $\tilde{K} \cong E_q^3 : \mathrm{GL}_3(q)$ is a normal 2-covering.

Assume next that q is even. Since $q+1$ is relatively prime to q, the non-identity unipotent elements of $\tilde{H}$ all lie in $\mathrm{SL}_2(q^2).2$. It can be verified that the Jordan forms of the non-identity unipotent elements of $\mathrm{SL}_2(q^2).2$ are

$$\begin{pmatrix} 1 & 1 & 0 & 0 \\ 0 & 1 & 0 & 0 \\ 0 & 0 & 1 & 1 \\ 0 & 0 & 0 & 1 \end{pmatrix}, \begin{pmatrix} 1 & 1 & 0 & 0 \\ 0 & 1 & 1 & 0 \\ 0 & 0 & 1 & 1 \\ 0 & 0 & 0 & 1 \end{pmatrix}.$$

Indeed, the Jordan form of the non-identity unipotent elements of $\mathrm{SL}_2(q)$, viewed as 2×2-matrices over $\mathbb{F}_{q^2}$, is

$$\begin{pmatrix} 1 & 1 \\ 0 & 1 \end{pmatrix}.$$

These matrices, viewed as 4×4-matrices over $\mathbb{F}_q$, have Jordan form consisting of two Jordan blocks of size 2. Now, let $\phi : \mathbb{F}_{q^2} \to \mathbb{F}_{q^2}$ be the Galois automorphism defined by $x \mapsto x^q$, $\forall x \in \mathbb{F}_{q^2}$. We now let $u\phi \in \mathrm{SL}_2(q^2).2$ be a 2-element and we study the Jordan form of $u\phi$, when viewed as an element of $\mathrm{SL}_4(q)$. Replacing u by a suitable $\mathrm{SL}_2(q^2)$-conjugate, we may suppose that

$$u = \begin{pmatrix} 1 & a \\ 0 & 1 \end{pmatrix},$$

for some $a \in \mathbb{F}_{q^2}$. We suppose that, for some $a \in \mathbb{F}_{q^2}$, the Jordan form of $u\phi$, viewed as an element of $\mathrm{SL}_4(q)$ does not have two Jordan blocks of size 2. Therefore, $u\phi$ has either two Jordan blocks of size 1 or $u\phi$ has a Jordan block of size at least 3. In the first, case $u\phi$ has order 2 and, in the second case, $u\phi$ has order 4. Observe that

$$(u\phi)^2 = uu^\phi = \begin{pmatrix} 1 & a + a^q \\ 0 & 1 \end{pmatrix}.$$

In particular, $(u\phi)^2$ viewed as a matrix of $\mathrm{SL}_4(q)$ is either the identity matrix or has two Jordan blocks of size 2. In the second case, $u\phi$ is a Jordan block of size

11.1 Linear Groups

4, as we claimed above. In the first case, $a + a^q = 0$, that is, $a \in \mathbb{F}_q$. Now, let $\varepsilon_1, \varepsilon_2$ be an $\mathbb{F}_{q^2}$-basis of $\mathbb{F}_{q^2}^2$ and let λ, λ^q be a Galois basis for $\mathbb{F}_{q^2}/\mathbb{F}_q$. Observe that $\lambda \varepsilon_1, \lambda^q \varepsilon_1, \lambda \varepsilon_2, \lambda^q \varepsilon_2$ is an $\mathbb{F}_q$-basis of $\mathbb{F}_{q^2}^2$. With respect to this basis the Jordan form of $u\phi$ is

$$\begin{pmatrix} 0 & 1 & 0 & a \\ 1 & 0 & a & 0 \\ 0 & 0 & 0 & 1 \\ 0 & 0 & 1 & 0 \end{pmatrix}.$$

This matrix has two Jordan blocks of size 2.

In particular, those Jordan forms have two blocks of size 2 or just one block of size 4. Note also that we have one more Jordan form than in the case of q odd. Thus, we need extra care to treat this case. Now, $\tilde{H}$ has no unipotent elements having two Jordan blocks of sizes 3 and 1. Therefore $\tilde{K}$ contains a unipotent element u having two Jordan blocks of sizes 3 and 1. Let $\tilde{P} \cong E_q^3$ be the Fitting subgroup of $\tilde{K}_M$. Using the subgroup structure of $\mathrm{GL}_3(q)$ it is not hard to verify that $\langle \tilde{P}, y, u \rangle = \tilde{K}_M$.

We deduce that either $\tilde{K} = \tilde{K}_M$ or $\tilde{K} \cong \mathrm{GL}_3(q)$ is a complement of $\tilde{P}$ in $\tilde{K}$. With [28], we deduce that all complements of $\tilde{P}$ in $\tilde{K}_M$ are conjugate. Summing up, either

$$\tilde{K} = \tilde{K}_M \cong E_q^3 : \mathrm{GL}_3(q) \text{ or } \mathrm{GL}_3(q) \cong \tilde{K} < \tilde{K}_M \cong E_q^3 : \mathrm{GL}_3(q),$$

where in the second possibility we may assume that $\tilde{K}$ is the standard complement of $\mathrm{GL}_3(q)$ in $E_q^3 : \mathrm{GL}_3(q)$.

We claim that $\mathrm{SL}_2(q^2).(q+1) \leq \tilde{H}$. Let w be an element of order $q^2 - 1$ in $\tilde{G} = \mathrm{SL}_4(q)$, having type $2 \oplus 2$ and such that $\mathbb{F}_q^4$ has only two distinct irreducible $\mathbb{F}_q\langle w \rangle$-submodules. To construct such an element w it suffices to take a Singer cycle s of $\mathrm{GL}_2(q)$ and consider $w = s \oplus (s^{-1})^T$. Now, w has no $\mathrm{Aut}(\tilde{G})$-conjugate in $\tilde{K}$ and hence w has an $\mathrm{Aut}(\tilde{G})$-conjugate in $\tilde{H}$. Without loss of generality, we may suppose that $w \in \tilde{H}$. Using the subgroup structure of $\mathrm{SL}_2(q^2)$ it is not hard to verify that $\langle x^{q+1}, w \rangle \cong \mathrm{SL}_2(q^2)$. Therefore, $\tilde{H} \geq \langle x, w \rangle \cong \mathrm{SL}_2(q^2).(q+1)$.

We claim that $\tilde{K} = \tilde{K}_M \cong E_q^3 : \mathrm{GL}_3(q)$. We argue by contradiction and we suppose that $\tilde{K} < \tilde{K}_M$ and hence, from what we have seen before, $\tilde{K} \cong \mathrm{GL}_3(q)$. Now, let

$$A = \begin{pmatrix} a_{11} & a_{12} \\ a_{21} & a_{22} \end{pmatrix} \in \mathrm{GL}_2(q)$$

be a Singer cycle of order $q^2 - 1$ and let $a := \det(A) \in \mathbb{F}_q$. From (2.3), a is a generator of the multiplicative group $\mathbb{F}_q^*$ and hence $o(a) = q - 1$. Since q is even, there exists $b \in \mathbb{F}_q$ with $b^2 = a^{-1}$. Let

$$g := \begin{pmatrix} b & 1 & 0 & 0 \\ 0 & b & 0 & 0 \\ 0 & 0 & a_{11} & a_{12} \\ 0 & 0 & a_{21} & a_{22} \end{pmatrix} \in \mathrm{SL}_4(q).$$

It is easy to verify that g has no $\mathrm{Aut}(\tilde{G})$-conjugate in $\tilde{K} \cong \mathrm{GL}_3(q)$. Therefore, g has an $\mathrm{Aut}(\tilde{G})$-conjugate in $\tilde{H} \leq \tilde{H}_M \cong \mathrm{SL}_2(q).(q+1).2$. Replacing g by a suitable $\mathrm{Aut}(\tilde{G})$-conjugate, we may suppose that $g \in \tilde{H}_M \cong \mathrm{SL}_2(q^2).(q+1).2 = (\mathrm{GL}_2(q^2) \cap \mathrm{SL}_4(q)).2$. Now, $o(g) = 2(q^2 - 1)$ and

$$g^{q^2-1} = \begin{pmatrix} 1 & 1 & 0 & 0 \\ 0 & 1 & 0 & 0 \\ 0 & 0 & 1 & 0 \\ 0 & 0 & 0 & 1 \end{pmatrix} \in \tilde{H}_M.$$

As we have seen above, g^{q^2-1} is not the Jordan form of an element in $(\mathrm{GL}_2(q^2) \cap \mathrm{SL}_4(q)).2$: the Jordan form of the non-identity unipotent elements in this group consists of either two Jordan blocks of size 2 or a single Jordan block of size 4. This shows that $\tilde{K}$ cannot be isomorphic to $\mathrm{GL}_3(q)$.

Routine computations with the possible Jordan forms of the elements of $\mathrm{SL}_4(q)$ show that $\tilde{H} \cong \mathrm{SL}_2(q^2).(q+1)$ and $\tilde{K} \cong E_q^3 : \mathrm{GL}_3(q)$ is a normal 2-covering.

11.2 Unitary Groups: Cases (8) and (9)

Suppose $G = \mathrm{PSU}_3(q)$ and we are in Case (8). Observe that $G \cong \tilde{G}$, because q is a power of 3 and $\gcd(3, q+1) = 1$. Here, we have $H \leq H_M \cong (q^2 - q + 1) : 3$ and $K \leq K_M \cong \mathrm{GU}_2(q)$. Now, G contains an element x of order $q^2 - q + 1$ and an element y of order $q^2 - 1$. As K_M contains no element having order $q^2 - q + 1$ and H_M contains no element having order $q^2 - 1$, up to replacing x and y with suitable $\mathrm{Aut}(G)$-conjugates, we deduce

$$C_{q^2-q+1} \cong \langle x \rangle \leq H \leq H_M = \mathbf{N}_G(\langle x \rangle) \cong (q^2 - q + 1) : 3$$

and

$$C_{q^2-1} \cong \langle y \rangle \leq K \leq K_M \cong \mathrm{SU}_2(q).(q+1).$$

Observe that $\langle x \rangle$ is a maximal subgroup of H_M. In particular, $H \in \{\langle x \rangle, \mathbf{N}_G(\langle x \rangle)\}$.

11.2 Unitary Groups

Assume $H = \langle x \rangle \cong C_{q^2-q+1}$. Then all the unipotent elements of G are $\mathrm{Aut}(G)$-conjugate to an element of K, however this is impossible because K_M contains no regular unipotent elements. Thus $H = H_M = \mathbf{N}_G(\langle x \rangle)$. The non-identity unipotent elements of $H = H_M$ are regular, and hence H contains no non-identity non-regular unipotent elements. Therefore, K contains at least one of these elements, say u. In particular 3 divides $|K|$. Now, $|\mathrm{GU}_2(q) : \mathrm{SU}_2(q)| = q + 1$, $\mathrm{GU}_2(q)/\mathrm{SU}_2(q)$ is cyclic of order $q + 1$ and $\mathrm{SU}_2(q) \cong \mathrm{SL}_2(q)$. Using the fact that K contains an element of order $q^2 - 1$ and an element of order 3 and using the subgroup structure of $\mathrm{SL}_2(q)$, we deduce that either $K = K_M$ or $K = P.(q+1)$, where $P \cong \langle y^{q+1}, u \rangle$ is a parabolic subgroup of $\mathrm{SL}_2(q)$. Recalling (11.1), we obtain $K \cong P.(q + 1)$. In particular, $|K| = q(q^2 - 1)$. However, in this case, we run into trouble with the elements of order $q + 1$. Indeed, as $q > 2$, G has two types of elements of order $q + 1$: elements having 2 eigenvalues in $\mathbb{F}_{q^2}$ and elements having 3 distinct eigenvalues in $\mathbb{F}_{q^2}$, see the proof of Lemma 4.2. It can be verified that the group K has no elements of the second type; therefore we have no more additional weak normal 2-coverings.

Suppose $G = \mathrm{PSU}_4(q)$ and we are in Case (9). Here it is more convenient to work with $\tilde{G} = \mathrm{SU}_4(q)$. We have $\tilde{H} \le \tilde{H}_M \cong \mathrm{GU}_3(q)$ and $\tilde{K} \le \tilde{K}_M \cong E_q^4 : \mathrm{SL}_2(q^2) : (q-1)$ is parabolic. Now, $\tilde{G}$ contains an element x of order $q^3 + 1$ and an element y of order $(q^4 - 1)/(q + 1)$. Using the fact that $q \ge 4$ we deduce that $\tilde{K}_M$ contains no element having order $q^3 + 1$ and $\tilde{H}_M$ contains no element having order $(q^4 - 1)/(q + 1)$, up to replacing x and y with suitable $\mathrm{Aut}(\tilde{G})$-conjugates, we deduce

$$C_{q^3+1} \cong \langle x \rangle \le \tilde{H} \le \tilde{H}_M \cong \mathrm{GU}_3(q)$$

and

$$C_{(q^4-1)/(q+1)} \cong \langle y \rangle \le \tilde{K} \le \tilde{K}_M \cong E_q^4 : \mathrm{SL}_2(q^2) : (q - 1).$$

Now let $z \in \tilde{G}$ having order $q^2 - 1$ such that the action of z on $V = \mathbb{F}_{q^2}^4$ is of type $2 \oplus 2$.[4] Since $\mathrm{GU}_3(q)$ contains no element $\mathrm{Aut}(\tilde{G})$-conjugate to z, we deduce that $\tilde{H}$ contains an element $\mathrm{Aut}(\tilde{G})$-conjugate to z. In particular, without loss of generality, we may assume $z \in \tilde{K}$.

We claim that

$$\tilde{K} = \tilde{K}_M.$$

---

[4] Observe that, for every $g \in \tilde{G}$ and for every $\varphi \in \mathrm{Aut}(\tilde{G})$, the elements g and g^φ have the same Jordan form.

As $\tilde{H} \leq \tilde{H}_M \cong \mathrm{GU}_3(q)$, $\tilde{H}$ contains no regular unipotent elements. Therefore, $\tilde{K}$ contains a regular unipotent element u. Let $\tilde{P}$ be the Fitting subgroup of $\tilde{K}_M$. Observe that $\mathbf{o}(y^{q-1}) = q^2 + 1$ is the order of a Singer cycle in $\mathrm{SL}_2(q^2)$ so that, by Proposition 2.5, it is indeed a Singer cycle of $\mathrm{SL}_2(q^2)$. Using the subgroup structure of $\mathrm{SL}_2(q^2)$ (see [9, Tables 8.1 and 8.2, p. 377]), it can be verified that either $\langle \tilde{P}, u, y^{q-1}, z^{q-1} \rangle / \tilde{P} \cong \mathrm{SL}_2(q^2)$ or $\langle \tilde{P}, u, y^{q-1}, z^{q-1} \rangle / \tilde{P} \cong \langle y^{q-1} \rangle.2$ and q is even. We show that the second possibility is ridiculous. Indeed, assume $\langle \tilde{P}, u, y^{q-1}, z^{q-1} \rangle / \tilde{P} \cong \langle y^{q-1} \rangle.2$ and q is even. As $\mathbf{o}(z^{q-1}) = q + 1$, $\mathbf{o}(y^{q-1}) = q^2 + 1$ and $\gcd(q + 1, q^2 + 1) = 1$, we deduce $z^{q-1} \in \tilde{P}$. As the order of z is relatively prime to the order of $\tilde{P}$, we deduce that $q^2 - 1 = \mathbf{o}(z) = q + 1$, which implies $q = 2$. However, this contradicts the fact that $q \geq 4$.

Thus $\tilde{K}\tilde{P} \geq \langle \tilde{P}, u, y^{q-1} \rangle \cong E_q^4 : \mathrm{SL}_2(q^2)$. As $y \in \tilde{K}$, we have $\tilde{K}\tilde{P} = \tilde{K}_M$. In particular, $\tilde{K}$ acts by conjugation irreducibly on $\tilde{P}$. Therefore, either $\tilde{K} = \tilde{K}_M$ or $\tilde{K} \cap \tilde{P} = 1$. In the first case, our claim is proved. In the second case, $\tilde{K} \cong \mathrm{SL}_2(q^2) : (q-1)$ is a complement of $\tilde{P}$ in $\tilde{K}_M$. By [28, Theorem (4.2) and Table (4.5)], $\tilde{K}$ is conjugate to the standard complement of $\mathrm{SL}_2(q^2) : (q-1)$ in $E_q^4 : \mathrm{SL}_2(q^2) : (q-1)$. However, we obtain a contradiction because this group contains no regular unipotent elements.

We now claim that

$$\tilde{H} = \tilde{H}_M$$

and hence (11.1) is not satisfied. We use the subgroup structure of $\mathrm{SU}_3(q)$ and the fact that $\tilde{H} \cap \mathrm{SU}_3(q)$ contains the Singer cycle x^{q+1}. Now, with the restriction that $q \geq 4$, the overgroups of $\langle x^{q+1} \rangle$ in $\mathrm{SU}_3(q)$ consist of only three groups

$$\langle x^{q+1} \rangle < \mathbf{N}_{\mathrm{SU}_3(q)}(\langle x^{q+1} \rangle) \cong (q^2 - q + 1).3 < \mathrm{SU}_3(q).$$

Therefore, $\tilde{H} = \langle x \rangle$, or $\tilde{H} = \mathbf{N}_{\tilde{G}}(\langle x \rangle)$, or $\tilde{H} = \tilde{H}_M$. However, the last possibility is excluded by (11.1) and by the fact that $\tilde{K} = \tilde{K}_M$. To conclude our analysis and reach a contradiction we look at elements having order $q + 1$. The elements of order $q + 1$ in $\tilde{G} = \mathrm{SU}_4(q)$ are of two types:

- they have 4 distinct eigenvalues in $\mathbb{F}_{q^2}$, or
- they have 3 (or less) distinct eigenvalues in $\mathbb{F}_{q^2}$.

The elements of order $q + 1$ in $\tilde{H}$ when $\tilde{H} = \langle y \rangle$ or $\tilde{H} = \mathbf{N}_{\tilde{G}}(\langle y \rangle)$ cannot have four distinct eigenvalues because these elements arise as the $q^2 - q + 1$ power of x. Similarly, the elements of order $q + 1$ in $\tilde{K} = E_q^4 : \mathrm{SL}_2(q^2) : (q - 1)$ cannot have four distinct eigenvalues.

11.3 Exceptional Groups　　　　　　　　　　　　　　　　　　　　　　　　　151

Summing up, when $\tilde{G} = \mathrm{SU}_4(q)$, the only weak normal 2-covering has maximal components.

11.3 Exceptional Groups: Case (2)

Suppose we are in Case (2), that is, $G = F_4(q)$, $H_M \cong {}^3D_4(q).3$, $K_M \cong \mathrm{Spin}_9(q) \cong 2.\Omega_9(q)$, $q = 3^a$ and $a \geq 1$. When $q = 3$, we have verified that the only weak normal 2-covering of G uses maximal components. Assume now $q > 3$. Now, G contains elements of order $q^4 - q^2 + 1$ and elements of order $(q+1)(q^3-1)$.[5] Let $x, y \in G$ with $\mathbf{o}(x) = q^4 - q^2 + 1$ and $\mathbf{o}(y) = (q+1)(q^3-1)$. As K_M contains no element having order $q^4 - q^2 + 1$ and H_M contains no element having order $(q+1)(q^3-1)$, we deduce that H contains elements of order $q^4 - q^2 + 1$ and K contains elements of order $(q+1)(q^3-1)$. Hence $(q^4 - q^2 + 1) \leq H \leq H_M \cong {}^3D_4(q).3$ and $(q+1)(q^3-1) \leq K \leq K_M \cong 2.\Omega_9(q)$.

We use the subgroup structure of ${}^3D_4(q).3$, see [9, Table 8.51, p. 404]. The overgroups of $\langle x \rangle$ in ${}^3D_4(q)$ consist of only four groups

$$\langle x \rangle < (q^4 - q^2 + 1).2 < \mathbf{N}_{{}^3D_4(q)}(\langle x \rangle) \cong (q^4 - q^2 + 1).[4] < {}^3D_4(q).$$

In particular, $H \cap {}^3D_4(q)$ is one of these groups. Now, the unipotent elements of $2.\Omega_9(q)$ have order $1, 3, 9$ and the unipotent elements of ${}^3D_4(q)$ have also order $1, 3, 9$.[6] However, in ${}^3D_4(q).3 \setminus {}^3D_4(q)$ there are unipotent elements having order 27.[7] This shows that H contains a unipotent element u having order 27. Thus $u^3 \in {}^3D_4(q)$ has order 9 and $\langle x, u^3 \rangle \leq H$. Therefore, $\langle x, u^3 \rangle \cong {}^3D_4(q)$ and $H_M = H = \langle x, u \rangle = {}^3D_4(q).3$.

We use the subgroup structure of $2.\Omega_9(q)$, see [9, Tables 8.58 and 8.59, p. 408]. Now, $2.\Omega_9(q)$ contains an element z of order $q^2 + 1$. As $H = H_M$ contains no element having order $q^2 + 1$, we deduce that K contains elements of order $q^2 + 1$. In particular, we may suppose, replacing z if necessary, that $z \in K$ and hence $\langle y, z \rangle \leq K$. In particular, $|K|$ is divisible by $(q^2+1)(q^2-1)(q^3-1)/2$. Using the subgroup structure of $2.\Omega_9(q)$ we deduce that $K = K_M \cong 2.\Omega_9(q)$.

Summing up, when $G = F_4(q)$, the only weak normal 2-covering has maximal components.

---

[5] This can be seen by the fact that $2.\Omega_9(q)$ contains elements of order $(q+1)(q^3-1)$ and ${}^3D_4(q)$ contains elements of order $q^4 - q^2 + 1$.

[6] This can be seen by the embedding of ${}^3D_4(q)$ in $\Omega_8^+(q^3)$, see for instance [9, p. 403], and by the fact that $\Omega_8^+(q^3)$ has no elements of order 27.

[7] We have verified that ${}^3D_4(q).3 \setminus {}^3D_4(q)$ contains elements having order 27 with a computer computation; indeed, we have checked that ${}^3D_4(3).3$ contains elements of order 27 and hence so does ${}^3D_4(q).3$ for any $q = 3^a$.

11.4 Exceptional Groups: Case (1)

Suppose now that we are in Case (1). Here, $G = G_2(q)$, $q \geq 4$ and q is even. When $q = 4$, we have used a computer to establish the result. Therefore, for the rest of the argument we suppose that $q \geq 8$. Here, we have $H \leq H_M \cong \mathrm{SL}_3(q).2$ and $K \leq K_M \cong \mathrm{SU}_3(q).2$. Now, H_M contains elements having order q^2+q+1 and K_M contains elements of order q^2-q+1. As K_M contains no element having order q^2+q+1 and H_M contains no element having order q^2-q+1, we deduce that H contains elements of order q^2+q+1 and K contains elements of order q^2-q+1. Hence $(q^2+q+1) \leq H \leq H_M \cong \mathrm{SL}_3(q).2$ and $(q^2-q+1) \leq K \leq K_M \cong \mathrm{SU}_3(q).2$.

Here we need the following remark, which follows from a direct computation.

Lemma 11.3 *Let $q \geq 4$ be even. If x, y are elements of order 8 in $\mathrm{SL}_3(q).2$, then x^2 and y^2 are $\mathrm{SL}_3(q)$-conjugate. Similarly, if x, y are elements of order 8 in $\mathrm{SU}_3(q).2$, then x^2 and y^2 are $\mathrm{SU}_3(q)$-conjugate.*

Using [27, Section 3.3] or [34, Section 3], we see that there exists a 1 to 1 correspondence between the conjugacy classes of maximal tori in $G_2(q)$ and the conjugacy classes in the Weyl group W of $G_2(q)$. Recall $W \cong D_{12}$ is dihedral of order 12. Under this correspondence, we see from [34, p. 507] that the two distinct conjugacy classes of reflections correspond to two $G_2(q)$-conjugacy classes of cyclic maximal tori of order $q^2 - 1$. Let $T_1 = \langle y_1 \rangle$ and $T_2 = \langle y_2 \rangle$ be two representatives of the $G_2(q)$-conjugacy classes of maximal tori of $G_2(q)$ of order $q^2 - 1$.

Now, both $\mathrm{SL}_3(q)$ and $\mathrm{SU}_3(q)$ contain a unique conjugacy class of maximal tori of order $q^2 - 1$. As T_1 and T_2 are not $G_2(q)$-conjugate, this shows that, interchanging T_1 and T_2 if necessary, we may suppose that $T_1 \leq \mathrm{SL}_3(q) \leq H_M$ and $T_2 \leq \mathrm{SU}_3(q) \leq K_M$.

We claim that T_1 and T_2 are in distinct $\mathrm{Aut}(G_2(q))$-conjugacy classes. Arguing by contradiction, we suppose that T_1 and T_2 are $\mathrm{Aut}(G_2(q))$-conjugate. From [29, Table 5, p. xvi], we have $\mathrm{Aut}(G_2(q)) = G_2(q) \rtimes \langle \phi \rangle$, where ϕ is a field automorphism. In particular, there exists $g \in G_2(q)$ and $i \in \mathbb{N}$ with $T_1^{g\phi^i} = T_2$. From [9, Table 8.30], $G_2(q)$ has a unique $G_2(q)$-conjugacy class of subgroups isomorphic to $\mathrm{SL}_3(q).2$. Therefore, as $\mathrm{SL}_3(q).2 \cong H_M \cong H_M^{g\phi^i}$, we deduce that there exists $h \in G_2(q)$ with $H_M^{g\phi^i h} = H_M$. Therefore, $T_2^h = (T_1^{g\phi^i})^h \leq H_M^{g\phi^i h} = H_M$. Since any two maximal tori of order $q^2 - 1$ of $\mathrm{SL}_3(q)$ are $\mathrm{SL}_3(q)$-conjugate, we deduce that there exists $k \in \mathrm{SL}_3(q) \leq H_M$ with $T_2^{hk} = T_1$, contradicting the fact that T_1 and T_2 are in distinct $G_2(q)$-conjugacy classes.

The previous paragraphs have shown that, if $z_1 \in \mathrm{SL}_3(q).2$ has order $q^2 - 1$ and $z_2 \in \mathrm{SU}_3(q).2$ has order $q^2 - 1$, then $\langle z_1 \rangle$ and $\langle z_2 \rangle$ are not $\mathrm{Aut}(G)$-conjugate. In particular, both H and K contain an element having order $q^2 - 1$. This yields $\mathrm{SL}_3(q) \leq H \leq H_M \cong \mathrm{SL}_3(q).2$ and $\mathrm{SU}_3(q) \leq K \leq K_M \cong \mathrm{SU}_3(q).2$.

11.5 Symplectic Groups

In the process of computing the character table for $G_2(q)$, Enomoto and Yamada [36, Section 1] have shown that there exists two conjugacy classes of cyclic unipotent subgroups of order 8. This can also be seen from [34, Table 1, p. 511], where Enomoto has shown that $G_2(q)$ contains 8 conjugacy classes of 2-elements: one class for the identity element, two classes with representatives x_3 and x_4 for the involutions, three classes with representatives x_5, x_6, x_7 for the elements of order 4 and two classes with representatives x_1, x_2 for the elements of order 8. Moreover, using the Chevalley relations, from [34, Table 1], we see that x_1^2 and x_2^2 are in distinct $G_2(q)$-conjugacy classes, because they are conjugate to x_5 and x_6 respectively.

Let $X \in \{H, K\}$. Suppose that x_1 and x_2 have both an $\mathrm{Aut}(G_2(q))$-conjugate in X. For $i \in \{1, 2\}$, let $x_i' \in X$ with x_i' $\mathrm{Aut}(G_2(q))$-conjugate to x_i. Thus, when $X = H$, $X = \langle \mathrm{SL}_3(q), x_1', x_2' \rangle = H_M \cong \mathrm{SL}_3(q).2$ and, when $X = K$, $X = \langle \mathrm{SU}_3(q), x_1', x_2' \rangle = K_M \cong \mathrm{SU}_3(q).2$. Now, x_1' and x_2' are two elements having order 8, therefore, by Lemma 11.3, $x_1'^2$ and $x_2'^2$ are $\mathrm{SL}_3(q).2$-conjugate or $\mathrm{SU}_3(q).2$-conjugate, contradicting the fact that $x_1'^2$ and $x_2'^2$ are in distinct $G_2(q)$-conjugacy classes.

The previous paragraph has shown that both H and K contain an element having order 8, coming from distinct $G_2(q)$-conjugacy classes. Thus $H = H_M \cong \mathrm{SL}_3(q).2$ and $K = K_M \cong \mathrm{SU}_3(q).2$.

11.5 Symplectic Groups: Cases (10) and (11)

We now consider Case (10), with $n \geq 6$. When $q \in \{2, 4\}$ and $n \in \{6, 8\}$, we have determined the weak normal 2-coverings with the help of a computer. In particular, when $n \in \{6, 8\}$, we may suppose that $q \geq 8$.

Here, we have $H \leq H_M \cong \mathrm{SO}_n^-(q)$ and $K \leq K_M \cong \mathrm{SO}_n^+(q)$. Now, for each $\ell \in \{0, \ldots, n/2 - 1\}$, G contains an element x_ℓ with

$$\mathrm{o}(x_\ell) = \begin{cases} \frac{(q^\ell - 1)(q^{n/2-\ell}+1)}{\gcd(q^\ell-1, q^{n/2-\ell}+1)} & \text{if } \ell > 0, \\ q^{n/2} + 1 & \text{if } \ell = 0. \end{cases}$$

As K_M contains no element $\mathrm{Aut}(G)$-conjugate to x_ℓ (one can see this for example using [50]), replacing x_ℓ with a suitable $\mathrm{Aut}(G)$-conjugate, we have $x_\ell \in H$ for all $\ell \in \{0, \ldots, n/2 - 1\}$. In particular, H contains the Singer cycle x_0 for $\Omega_n^-(q)$. By using Lemma 7.3 when $n \geq 8$ and [9] when $n = 6$, we obtain that one of the following holds

- there exists a prime s with s dividing $n/2$, $n/s \geq 4$ such that $H \leq \Omega_{n/s}^-(q^s).s$,
- $n/2$ is odd and $H \leq \mathrm{GU}_{n/2}(q).2$, or
- $\Omega_n^-(q) \leq H$.

Here we are using the fact that, when $n = 6$, $q \geq 8$. Now, it is not hard to check that it is impossible for $|\Omega_{n/s}^{-}(q^s).s|$ or for $|\mathrm{GU}_{n/2}(q).2|$ to be divisible by $(q^\ell - 1)(q^{n/2-\ell}+1)/\gcd(q^\ell - 1, q^{n/2-\ell}+1)$, for every $\ell \in \{1,\ldots,n/2-1\}$. Therefore,

$$\Omega_n^-(q) \leq H \leq \mathrm{SO}_n^-(q) \cong \Omega_n^-(q).2.$$

Now, for each $\ell \in \{1,\ldots, n/2 - 1\}$, G contains an element y_ℓ with

$$\mathrm{o}(y_\ell) = \frac{(q^\ell + 1)(q^{n/2-\ell}+1)}{\gcd(q^\ell + 1, q^{n/2-\ell}+1)}.$$

and type $2\ell \oplus (n - 2\ell)$. As H_M contains no element $\mathrm{Aut}(G)$-conjugate to y_ℓ, replacing y_ℓ with a suitable $\mathrm{Aut}(G)$-conjugate, we have $y_\ell \in K$ for all $\ell \in \{1,\ldots,n/2-1\}$. In particular, K contains y_1, which has type $2 \oplus (n-2)$. By using Lemma 8.1 when $n \geq 10$ and [9] when $n \in \{6, 8\}$, we obtain that one of the following holds

- $K \leq (\Omega_2^-(q) \perp \Omega_{n-2}^-(q)).[4]$, or
- $n/2$ is even and $K \leq \mathrm{GU}_{n/2}(q).[4]$, or
- $\Omega_n^+(q) \leq K$.

Here we are using the fact that, when $n \in \{6, 8\}$, $q \geq 8$. Now, it is not hard to check that it is impossible for $|\mathrm{GU}_{n/2}(q).[4]|$ to be divisible by $(q^i + 1)(q^{n/2-i}+1)/\gcd(q^i + 1, q^{n/2-i}+1)$, for every $i \in \{1,\ldots,n/2-1\}$. Similarly, $|(\Omega_2^-(q) \perp \Omega_{n-2}^-(q)).[4]|$ is divisible by $(q^i+1)(q^{n/2-i}+1)/\gcd(q^i+1, q^{n/2-i}+1)$, for every $i \in \{1,\ldots,n/2-1\}$, if and only if $q \in \{2,4\}$. However, when $n \geq 8$, the group $\Omega_2^-(q) \perp \Omega_{n-2}^-(q)$ cannot contain y_2, which has type $4 \oplus (n-4)$. Therefore,

$$\Omega_n^+(q) \leq K \leq \mathrm{SO}_n^+(q) \cong \Omega_n^+(q).2.$$

We summarize here Lemma 6.6. The group $G = \mathrm{Sp}_n(q)$ contains regular unipotent elements u_+ and u_-, with u_+ preserving a non-degenerate quadratic form having Witt index 0 and polarizing the symplectic form on G and with u_- preserving a non-degenerate quadratic form having Witt index 1 and polarizing the symplectic form on G. Furthermore, u_+ and u_- are in distinct $\mathrm{Aut}(\mathrm{Sp}_n(q))$-conjugacy classes. Now, $\Omega_n^+(q)$ and $\Omega_n^-(q)$ do not contain regular unipotent elements and $u_+ \in \mathrm{SO}_n^+(q) \setminus \Omega_n^+(q)$ and $u_- \in \mathrm{SO}_n^-(q) \setminus \Omega_n^-(q)$. Moreover, u_+ has no $\mathrm{Aut}(\mathrm{Sp}_n(q))$-conjugate in $\mathrm{SO}_n^-(q)$ and u_- has no $\mathrm{Aut}(\mathrm{Sp}_n(q))$-conjugate in $\mathrm{SO}_n^+(q)$. In particular, replacing u_+ and u_- by suitable $\mathrm{Aut}(G)$-conjugates, we have that $u_- \in H \setminus \Omega_n^-(q)$ and $u_+ \in K \setminus \Omega_n^+(q)$. Therefore, $H = H_M \cong \mathrm{SO}_n^-(q)$ and $K = K_M \cong \mathrm{SO}_n^+(q)$.

We now consider Case (10), with $n = 4$. When $q = 4$, we have determined the weak normal 2-coverings with the help of a computer. Therefore, for the rest of the proof, we suppose $q \geq 8$. Here, we have $H \leq H_M \cong \mathrm{SO}_4^-(q)$ and $K \leq K_M \cong \mathrm{SO}_4^+(q)$. The group G contains a Singer cycle x having order $q^2 + 1$ and,

11.5 Symplectic Groups

in the proof of Lemma 5.4, we have proved that G contains an element y having order $q+1$ and with $|C_G(y)| = (q+1)^2$. Moreover, we have shown that y has no Aut(G)-conjugate in $H_M \cong \mathrm{SO}_4^-(q)$. Therefore,

$$C_{q^2+1} \cong \langle x \rangle \leq H \leq H_M \cong \mathrm{SO}_4^-(q) \text{ and } C_{q+1} \cong \langle y \rangle \leq K \leq K_M \cong \mathrm{SO}_4^+(q).$$

Now, $\mathrm{SO}_4^-(q) \cong \mathrm{SL}_2(q^2).2$ and $\mathrm{SO}_4^+(q) \cong (\mathrm{SL}_2(q) \times \mathrm{SL}_2(q)).2$. By consulting the maximal subgroups of $\mathrm{SL}_2(q^2)$, we see that one of the following holds:

- $H \leq \mathbf{N}_{\mathrm{SO}_4^-(q)}(\langle x \rangle) \cong (q^2+1).[4]$, or
- $\Omega_4^-(q) \leq H$.

Similarly, by consulting the maximal subgroups of $\mathrm{SL}_2(q) \times \mathrm{SL}_2(q)$, we see that one of the following holds:

- $C_{q+1} \times C_{q+1} \leq K \leq (D_{2(q+1)} \times D_{2(q+1)}).2$, or
- $\Omega_4^+(q) \leq K$.

Using Lemma 6.6 with $n = 4$, we deduce (replacing u_+ and u_- with suitable conjugates if necessary) that $u_- \in H \setminus \Omega_n^-(q)$ and $u_+ \in K \setminus \Omega_n^+(q)$. Taking this into account, we have that one of the followings holds:

- $H = \mathbf{N}_{\mathrm{SO}_4^-(q)}(\langle x \rangle) \cong (q^2+1).[4]$, or
- $H = \mathrm{SO}_4^-(q)$.

Moreover,

- $(C_{q+1} \times C_{q+1}).4 \leq K \leq (D_{2(q+1)} \times D_{2(q+1)}).2$, or
- $K = \mathrm{SO}_4^+(q)$.

We claim that $H \cong (q^2+1).[4]$ and $K \cong \mathrm{SO}_4^+(q)$ does give rise to a weak normal 2-covering. Indeed, since $\mathrm{SO}_4^+(q)$ and $\mathrm{SO}_4^-(q)$ is a normal 2-covering, it suffices to show that each element g in $\mathrm{SO}_4^-(q) \setminus H$ has an Aut($\mathrm{Sp}_4(q)$)-conjugate in either H or $K \cong \mathrm{SO}_4^+(q)$. Using the fact that $\mathrm{SO}_4^-(q) \cong \mathrm{SL}_2(q^2) : 2$, we see that $\mathbf{o}(g)$ is a divisor of $q^2 + 1$, or $q^2 - 1$, or 4, or $2(q-1)$, or $2(q+1)$. Now, the argument is a case by case analysis using also the parametrization of the $\mathrm{Sp}_4(q)$-conjugacy classes in [35]. We just give details when $\mathbf{o}(g) = 2(q+1)$ and omit the other cases. Here, $g = su$, where $\mathbf{o}(u) = 2$ and $\mathbf{o}(s) = q+1$. In particular, s is a semisimple element and hence the vector space V decomposes as the direct sum of non-degenerate symplectic spaces $V_1 \oplus V_2$ with $\dim V_1 = \dim V_2 = 2$. Observe that u either fixes setwise V_1 and V_2 or u swaps V_1 and V_2. Now, a graph automorphism of $\mathrm{Sp}_4(q)$ maps the $\mathcal{C}_8$-subgroup $K = \mathrm{SO}_4^+(q)$ to a $\mathcal{C}_2$-subgroup $(\mathrm{Sp}_2(q) \times \mathrm{Sp}_2(q)) : 2$ and hence g has an Aut($\mathrm{Sp}_4(q)$)-conjugate in K.

We claim that the weak normal 2-covering $H \cong (q^2+1).[4]$ and $K \cong \mathrm{SO}_4^+(q)$ does not give rise to normal 2-coverings. We argue by contradiction. Here we need to be careful with the role played by the graph automorphism of $\mathrm{Sp}_4(q)$. We argue using elements having order $q^2 - 1$. As H contains no elements having order $q^2 - 1$, we deduce that each element having order $q^2 - 1$ in $\mathrm{Sp}_4(q)$ has an $\mathrm{Sp}_4(q)$-conjugate in K. Let $g := s \oplus s^{(-1)^T}$, where $s \in \mathrm{GL}_2(q)$ is a Singer cycle and g fixes a 2-dimensional totally isotropic subspace of $\mathbb{F}_q^4$. Since g has an $\mathrm{Sp}_4(q)$-conjugate in H, we deduce that K cannot be in the Aschbacher class $\mathcal{C}_2$. Thus K must be in the Aschbacher class $\mathcal{C}_8$, see [9, Table 8.15]. Now, we suppose that the matrix defining the symplectic form for $\mathrm{Sp}_4(q)$ is

$$J := \begin{pmatrix} 0 & 1 & 0 & 0 \\ 1 & 0 & 0 & 0 \\ 0 & 0 & 0 & 1 \\ 0 & 0 & 1 & 0 \end{pmatrix}.$$

Without loss of generality, we may suppose that $\mathrm{SO}_4^+(q)$ preserves the quadratic form

$$x_1 x_2 + x_3 x_4.$$

Let $A \in \mathrm{SL}_2(q)$ be a Singer cycle having order $q+1$ and let $a \in \mathbb{F}_q^*$ with $\mathbf{o}(a) = q-1$. Set

$$g := \begin{pmatrix} A & 0 & 0 \\ 0 & a & 0 \\ 0 & 0 & a^{-1} \end{pmatrix}.$$

Now, g induces on $\langle e_1, e_2 \rangle$ a matrix having order $q+1$ and hence the putative quadratic form of Witt index 0 preserved by g must restrict to $\langle e_1, e_2 \rangle$ to a quadratic form having Witt index 1, because $\mathrm{O}_2^+(q)$ has no elements of order $q+1$. However, this is impossible.

Recall that, by (11.1), we have $H < H_M$ or $K < K_M$. Clearly, when $H = \mathbf{N}_{\mathrm{SO}_4^-(q)}(\langle x \rangle) \cong (q^2+1).[4]$ and $(C_{q+1} \times C_{q+1}).4 \leq K \leq (D_{2(q+1)} \times D_{2(q+1)}).2$, we have no weak normal 2-coverings, because we are not covering elements having order $q-1$. Finally, suppose that $H \cong \mathrm{SO}_4^-(q)$ and $(C_{q+1} \times C_{q+1}).4 \leq K \leq (D_{2(q+1)} \times D_{2(q+1)}).2$. we argue by contradiction and we suppose that H and K do form a weak normal 2-covering. As K has no element of order $q-1$, we deduce that each element having order $q-1$ in $\mathrm{Sp}_4(q)$ has an $\mathrm{Aut}(\mathrm{Sp}_4(q))$-conjugate in H. We use the matrix J in the previous paragraph for the symplectic form on $\mathrm{Sp}_4(q)$. Let $a, b \in \mathbb{F}_q^*$ with $\mathbf{o}(a) = \mathbf{o}(b) = q-1$ and with $\{a, a^{-1}, b, b^{-1}\}$ having cardinality 4. Observe that this is possible because $q \geq 8$. Set

$$g := \begin{pmatrix} a & 0 & 0 & 0 \\ 0 & a^{-1} & 0 & 0 \\ 0 & 0 & b & 0 \\ 0 & 0 & 0 & b^{-1} \end{pmatrix}.$$

11.5 Symplectic Groups

Recall again that under a graph automorphism H is in the Aschbacher class $\mathcal{C}_3$ or $\mathcal{C}_8$. The elements of order $q - 1$ in the $\mathcal{C}_3$-class $\mathrm{SL}_2(q^2) : 2$ cannot have four distinct eigenvalues. So, g must be conjugate to an element of a maximal subgroup $\mathrm{SO}_4^-(q)$ within the Aschbacher class $\mathcal{C}_8$, either through an inner automorphism or a field automorphism. Now, $\langle e_1, e_2 \rangle$ and $\langle e_3, e_4 \rangle$ are two 2-dimensional non-degenerate subspaces of $\mathbb{F}_q^4$ which are mutually orthogonal. Moreover, on each subspace g induces a matrix of order $q - 1$. Therefore, our putative quadratic form having Witt index 1 must restrict to both $\langle e_1, e_2 \rangle$ and $\langle e_3, e_4 \rangle$ to a quadratic form having Witt index 1, because $\mathrm{O}_2^+(q)$ does not contain elements of order $q + 1$. Thus we have reached in this way a contradiction.

We now consider Case (11). Here it is more convenient to work with $\tilde{G} = \mathrm{Sp}_6(q)$. We have $\tilde{H} \leq \tilde{H}_M \cong \mathrm{Sp}_2 \perp \mathrm{Sp}_4(q)$ and $\tilde{K} \leq \tilde{K}_M \cong \mathrm{Sp}_2(q^3).3$. Now, $\tilde{G}$ contains an element x of order $q^2 + 1$ and an element y of order $q^3 + 1$. As $\tilde{K}_M$ contains no element having order $q^2 + 1$ and $\tilde{H}_M$ contains no element having order $q^3 + 1$, up to replacing x and y with suitable $\mathrm{Aut}(\tilde{G})$-conjugates, we deduce

$$C_{q^2+1} \cong \langle x \rangle \leq \tilde{H} \leq \tilde{H}_M \cong \mathrm{Sp}_2(q) \perp \mathrm{Sp}_4(q)$$

and

$$C_{q^3+1} \cong \langle y \rangle \leq \tilde{K} \leq \tilde{K}_M \cong \mathrm{Sp}_2(q^3).3.$$

We claim that $\tilde{K} = \tilde{K}_M$. The group $\tilde{G}$ contains unipotent elements u_1 having one Jordan block of size 6. Elements of this type are not in $\tilde{H}_M \cong \mathrm{Sp}_2(q) \perp \mathrm{Sp}_4(q)$ and hence u_1 has an $\mathrm{Aut}(\tilde{G})$-conjugate in $\tilde{K}$. Observe that u_1 has order 9 and hence $\langle y, u_1^3 \rangle$ is isomorphic to a subgroup of $\mathrm{Sp}_2(q^3)$ having order divisible by $3(q^3 + 1)$. Considering the subgroup structure of $\mathrm{Sp}_2(q^3)$, we deduce $\langle u_1^3, y \rangle \cong \mathrm{Sp}_2(q^3)$. Hence $\tilde{K} = \langle u_1, y \rangle = \tilde{K}_M$.

We claim that $\tilde{H} = \tilde{H}_M$, contradicting (11.1). Among the elements of order $q^2 + 1$ we can choose x fixing pointwise a 2-dimensional subspace of $\mathbb{F}_q^6$ and hence $x \in \mathrm{Sp}_4(q)$. The group $\tilde{G}$ contains unipotent elements u_2 having three Jordan blocks of size 1, 1 and 4 respectively. Elements of this type are not in $\tilde{K} = \tilde{K}_M$ because the non-identity unipotent elements in $\tilde{K}$ have Jordan blocks of type 6, or $2 \oplus 2 \oplus 2$, or $3 \oplus 3$. Hence u_2 has an $\mathrm{Aut}(\tilde{G})$-conjugate in $\tilde{H}$. Observe that u_2 has order 9. Hence $\langle x, u_2 \rangle$ is isomorphic to a subgroup of $\mathrm{Sp}_4(q)$ having order divisible by $9(q^2 + 1)$ and containing a cyclic subgroup of order 9. Considering the subgroup structure of $\mathrm{Sp}_4(q)$, we deduce $\langle u_2, x \rangle \cong \mathrm{Sp}_4(q)$. To show that the remaining $\mathrm{Sp}_2(q)$ direct factor of $\tilde{H}_M$ is also contained in $\tilde{H}$ it suffices to consider elements x' having order $(q^2+1)(q+1)/2$ and type $2 \oplus 4$ (where the matrix induced on the two invariant subspaces has order $q + 1$ and $q^2 + 1$, respectively) and u_2' having order 9 and consisting of two Jordan blocks of size 2, 4. These elements are not $\mathrm{Aut}(\tilde{G})$-conjugate to elements in $\tilde{K}$ and hence we may assume that $u_2', x' \in \tilde{H}$. Now, $\mathrm{Sp}_2(q) \perp \mathrm{Sp}_4(q) \cong \langle u_2, x, u_2', x' \rangle \leq \tilde{H}$.

11.6 Proof of Theorem 2 of Burness and Tong-Viet

In this section, we use Theorem 11.1 to prove a rather interesting result of Burness and Tong-Viet.

Let A be a transitive permutation group on a finite set Ω. We recall that, an element $x \in A$ is a derangement if it acts fixed-point-freely on Ω, that is, for every $\omega \in \Omega$, $\omega^x \ne \omega$. We write $\mathcal{D}(A)$ for the set of derangements in A. If H is a point stabilizer, then x is a derangement if and only if $x^A \cap H = \emptyset$, where x^A is the A-conjugacy class of x. Therefore,

$$A = \mathcal{D}(A) \cup \bigcup_{a \in A} H^a,$$

where this union is disjoint.

In [21], Burness and Tong-Viet are interested in transitive permutation groups A with the special property that, for some prime number r, every derangement is an r-element, that is, has order a power of r. In particular, if R is a Sylow r-subgroup of A, then

$$A = \bigcup_{a \in A} R^a \cup \bigcup_{a \in A} H^a.$$

The authors are interested only in primitive permutation groups and they show that, in this case, the socle of A is either abelian or non-abelian simple. Moreover, when the socle of A is non-abelian simple, the group A is almost simple and they give a complete classification of the possible choices for the primitive group A and for the prime r. We report here their main result for almost simple groups.

Theorem 11.4 *Let A be a finite almost simple primitive permutation group with point stabilizer H. Then every derangement in A is an r-element for some fixed prime r if and only if (A, H, r) is one of the cases in Table 11.10.*[8]

Proof From the paragraphs leading to the statement of Theorem 11.4, we see that given an almost simple primitive group A with point stabilizer H, every derangement in A is an r-element for a fixed prime r if and only if $\{H, R\}$ is a normal 2-covering of A, where R is a Sylow r-subgroup. Note that H is a proper subgroup of A, because H is core-free in A, and R is a proper subgroup of A, because A is not an r-group.

---

[8] We are reporting Table 11.10 using the notation in our work.

11.6 Proof of Theorem 2 of Burness and Tong-Viet

Table 11.10 Examples arising in Theorem 11.4

A	H	r	Conditions	Remarks
M_{11}	$PSL_2(11)$	2		
$PSL_2(8)$	P_1 or D_{14}	3		P_1 parabolic
$PSL_2(q)$	P_1	r	$q = 2r^e - 1$	P_1 parabolic
	P_1 or $D_{2(q-1)}$	r	$r = q + 1$ Fermat prime	P_1 parabolic
	$D_{2(q+1)}$	r	$r = q - 1$ Mersenne prime	
$PGL_2(q)$	P_1	2	$q = 2^{e+1} - 1$ Mersenne prime	P_1 parabolic
$P\Gamma L_2(q)$	$\mathbf{N}_G(D_{2(q+1)})$	r	$r = q - 1$ Mersenne prime	
$P\Gamma L_2(8)$	P_1 or $\mathbf{N}_G(D_{14})$	3		P_1 parabolic
$PSL_3(q)$	P_1 or P_2	r	$q^2 + q + 1 = \gcd(3, q - 1)r$ or $q^2 + q + 1 = 3r^2$	P_1, P_2 parabolic

Let G be the socle of A and observe that G is a non-abelian simple group. Now, $H \cap G$ is a proper subgroup of G, because H is a core-free subgroup of A. Moreover, $R \cap G$ is a proper subgroup of G, because G is not an r-group. Furthermore,

$$G = G \cap A = G \cap \left(\bigcup_{a \in A} H^a \cup \bigcup_{a \in A} R^a \right) = \bigcup_{a \in A} (H \cap G)^a \cap \bigcup_{a \in A} (R \cap G)^a$$
$$= \bigcup_{a \in \mathrm{Aut}(G)} (H \cap G)^a \cap \bigcup_{a \in \mathrm{Aut}(G)} (R \cap G)^a$$

and hence $\{H \cap G, R \cap G\}$ is a weak normal 2-covering of G appearing in Theorem 11.1, with the extra information that $R \cap G$ is a Sylow subgroup of G.

Now the proof follows with a direct inspection of Tables 11.1, 11.2, 11.3, 11.4, 11.5, 11.6, 11.7, 11.8, and 11.9. We do not give details of this computations, but they are all straightforward. Observe that the inspection of Table 11.9 is via a computer computation, because we have omitted in Table 11.9 structural information on the two components of the weak normal 2-coverings. □

Chapter 12
Degenerate Normal 2-Coverings

In this chapter we aim to give some partial results concerning normal 2-coverings of almost simple groups which, in some sense, we consider ***degenerate***. Let A be an almost simple group with socle G and let $\mu = \{H, K\}$ be a normal 2-covering of A. We say that μ is degenerate if either $G \leq H$ or $G \leq K$. As we have discussed in Sect. 1.2, there are two cases to consider

DI: $\quad G \leq H \cap K$,
DII: $\quad$ interchanging H and K, if necessary, $H \geq G$ and $K \not\geq G$.

12.1 The Degenerate Normal 2-Coverings of the First Type

To discuss the degenerate normal 2-coverings of the first type we only need two preliminary remarks.

Lemma 12.1 *Let X be a finite nilpotent group. Then either X is cyclic and hence X has no normal coverings, or the normal covering number of X is not* 2.

Proof We argue by contradiction and we suppose that there exists a non-cyclic nilpotent group X having normal covering number 2. Let $\{V, W\}$ be a normal 2-covering of X. Without loss of generality, we may suppose that V and W are both maximal subgroups of X. As X is nilpotent, $V \trianglelefteq X$ and $W \trianglelefteq X$. Therefore V and W are normal proper subgroups of X with $X = V \cup W$, however this is impossible. □

Lemma 12.2 *Let G be a non-abelian simple group. Then the outer automorphism group of G is non-nilpotent if and only if G is isomorphic to one of the following groups:*

(1) $\mathrm{PSL}_n(q)$ *with* $\gcd(n, q-1)$ *divisible by an odd prime,*
(2) $\mathrm{PSU}_n(q)$ *with* $\gcd(n, q+1)$ *divisible by an odd prime,*
(3) $\mathrm{P}\Omega_8^+(q)$,
(4) $E_6(q)$ *with 3 dividing* $q-1$,
(5) $^2E_6(q)$ *with 3 dividing* $q+1$.

Proof The structure of the outer automorphism group of the non-abelian simple groups is in [66, Chapter 2] for classical groups, in [49, Section 2.5] for exceptional groups and, for instance in [29], for alternating and sporadic groups. Using these references, we deduce that the outer automorphism group of G is not nilpotent if and only if G is isomorphic to one of the groups listed in the statement of this theorem. □

Theorem 12.3 *Let A be an almost simple group with socle G and let* $\{H, K\}$ *be a normal 2-covering of A with* $G \leq H \cap K$. *Then* A/G *is not nilpotent and G is isomorphic to one of the groups in (1)– (5) of Lemma 12.2.*

Conversely, for each of these five cases, $\mathrm{Aut}(G)$ *admits a normal 2-covering* $\{H, K\}$ *with* $G \leq H \cap K$, *that is,* $\mathrm{Aut}(G)$ *admits a degenerate normal 2-covering of the first type.*

Proof Let $X := A/G$, $V := H/G$ and $W := K/G$. Since V and W are proper subgroups of X, we have that $\{V, W\}$ are the components of a normal 2-covering of X. It follows that X is not cyclic. Thus Lemma 12.1 implies that $X = A/G$ is not nilpotent. As a consequence $\mathrm{Aut}(G)/G$ is not nilpotent and Lemma 12.2 implies that G is in one of the five cases (1)– (5) of Lemma 12.2. Suppose now that G is in one of the five cases (1)– (5) of Lemma 12.2. Using the same references appearing in the proof of Lemma 12.2, it can be verified that $\mathrm{Aut}(G)$ admits a normal subgroup $N \geq G$ with $\mathrm{Aut}(G)/N$ isomorphic to a dihedral group D of order $2p$, for some odd prime number p. For instance, when $G := E_6(q)$, using [49, Theorem 2.5.12 (i)], we see that $\mathrm{Aut}(G)/G$ has a quotient isomorphic to the dihedral group of order $2 \cdot 3$. As D has normal covering number 2, we deduce that $\mathrm{Aut}(G)$ has also normal covering number 2 and admits a degenerate normal 2-covering of the first type. □

12.2 The Degenerate Normal 2-Coverings of the Second Type

We now deal with the degenerate normal 2-coverings of almost simple groups of the second type. A key tool to do this is the theory of **Shintani descent**, which we outline here following the work of Kawanaka [64, Section 2]. We also refer to [20, Section 2] and the more recent improvements of Harper on the theory of Shintani descent [56].

12.2 The Degenerate Normal 2-Coverings of the Second Type

Let X be a connected linear algebraic group defined over an algebraically closed field and let $\sigma : X \to X$ be a Frobenius morphism.[1] Thus, σ is a bijective endomorphism of algebraic groups with finite fixed point subgroup

$$X_\sigma := \{x \in X \mid x^\sigma = x\}.$$

Let e be a positive integer and set $\mathcal{G} := X_{\sigma^e}$ and $\mathcal{H} := X_\sigma \leq \mathcal{G} \leq X$. As $\mathcal{G}$ is σ-stable, the restriction σ' of $\sigma : X \to X$ to $\mathcal{G}$ is an automorphism of $\mathcal{G}$ having order e. By abuse of notation, we identify σ' with σ when considering the action of σ on $\mathcal{G}$. Let $\mathcal{A} := \langle \sigma \rangle$ and consider the semidirect product $\mathcal{G} \rtimes \mathcal{A}$ with multiplication

$$\sigma^i s \cdot \sigma^j t = \sigma^{i+j} s^{\sigma^j} t,$$

for every $i, j \in \mathbb{Z}$ and $s, t \in \mathcal{G}$.

Let $s \in \mathcal{G}$ and consider the element σs in the coset $\sigma \mathcal{G} = \mathcal{G}\sigma$ of $\mathcal{G} \rtimes \mathcal{A}$. Then we have $(\sigma s)^2 = \sigma^2 s^\sigma s$ and an inductive argument gives

$$(\sigma s)^e = s^{\sigma^{e-1}} s^{\sigma^{e-2}} \cdots s^\sigma s \in \mathcal{G}. \tag{12.1}$$

By the Lang-Steinberg theorem [49, Theorem 2.1.1], there exists $a \in X$ with $s = a^{-\sigma} a$. Using (12.1), we find $a(\sigma s)^e a^{-1} \in X_\sigma = \mathcal{H}$. Indeed,

$$(a(\sigma s)^e a^{-1})^\sigma = (a(s^{\sigma^{e-1}} s^{\sigma^{e-2}} \cdots s^\sigma s) a^{-1})^\sigma = a^\sigma s^{\sigma^e} s^{\sigma^{e-1}} \cdots s^{\sigma^2} s^\sigma a^{-\sigma}$$

$$= a s^{-1} s^{\sigma^e} s^{\sigma^{e-1}} \cdots s^{\sigma^2} s^\sigma s a^{-1} = a s^{\sigma^{e-1}} \cdots s^{\sigma^2} s^\sigma s a^{-1}$$

$$= a(\sigma s)^e a^{-1},$$

where in the fourth equality we used the fact that $s^{\sigma^e} = s$. Therefore, $a(\sigma s)^e a^{-1} \in \mathcal{H}$.

This observation is the key remark for setting up the Shintani correspondence. Indeed, the Shintani correspondence, or Shintani descent, is a map f from the $\mathcal{G} \rtimes \mathcal{A}$-conjugacy classes in the coset $\sigma \mathcal{G}$ to the set of $\mathcal{H}$-conjugacy classes in $\mathcal{H}$. The mapping f is defined by

$$f : (\sigma s)^{\mathcal{G} \rtimes \mathcal{A}} \mapsto (a(\sigma s)^e a^{-1})^\mathcal{H}.$$

We summarize the basic properties of this mapping in the following lemma, see for instance [20, Lemma 2.13].

---

[1] There is a slight abuse of terminology here with the term "Frobenius morphism"; however, we use this term since we are following [20, Section 2].

Lemma 12.4 *The mapping f is a bijection between the $\mathcal{G} \rtimes \mathcal{A}$-conjugacy classes in the coset σG and the set of H-conjugacy classes. Moreover,*

$$\mathbf{C}_G(\sigma s) = a^{-1}\mathbf{C}_{\mathcal{H}}(f(\sigma s))a = \mathbf{C}_{a^{-1}\mathcal{H}a}((\sigma s)^e).$$

Theorem 12.5 *Let A be an almost simple group with socle G and let $\{H, K\}$ be a normal 2-covering of A with $G \leq H$ and $G \not\leq K$. Let H_M be a maximal subgroup of A with $H \leq H_M$ and K_M a maximal subgroup of A with $K \leq K_M$. Then one of the following holds:*

(1) *A/G is not nilpotent and G is isomorphic to one of the groups in (1)–(5) of Lemma 12.2;*
(2) *A/G is nilpotent and one of the following holds:*

  (a) *G is a group of Lie type and $K_M \cap G$ is the centralizer in G of a field automorphism of odd prime order r. Moreover, r is not the characteristic of G, unless $G = \mathrm{PSL}_2(q)$;*
  (b) *$G = \mathrm{PSL}_2(2^f)$, $f > 1$ is odd, $A = \mathrm{Aut}(\mathrm{PSL}_2(2^f)) = \mathrm{PSL}_2(2^f) \rtimes \langle \phi \rangle$, $H = H_M = \mathrm{PSL}_2(2^f) \rtimes \langle \phi^p \rangle$ for some prime divisor p of f is normal in A and $K_M \cong D_{2(2^f+1)}.\langle \phi \rangle$, where ϕ is a field automorphism of order f;*
  (c) *$G = \mathrm{PSL}_2(p^f)$, p is an odd prime number, f is an even natural number, $A = G\langle \iota \phi^i \rangle$, $H = H_M = G \rtimes \langle \phi^{2i} \rangle$ has index 2 in A, $\iota \in \mathrm{PGL}_2(p^f) \setminus \mathrm{PSL}_2(p^f)$, ϕ is a field automorphism of $\mathrm{PSL}_2(p^f)$ of order f, i is a divisor of f with f/i even and $K_M \cap G = D_{p^f-1}$;*
  (d) *$G = \mathrm{Sz}(2^f)$, $A = \mathrm{Aut}(\mathrm{Sz}(2^f)) = \mathrm{Sz}(2^f) \rtimes \langle \phi \rangle$, $H = \mathrm{Sz}(2^f) \rtimes \langle \phi^p \rangle$ for some prime divisor p of f and $K_M = \mathbf{N}_A(P) \cong (q \pm \sqrt{2q} + 1) : 4 : f$, where P is a Sylow 5-subgroup of G and ϕ is a field automorphism of order f;[2]*
  (e) *$G = \mathrm{PSU}_3(2^f)$ with $f > 1$ odd, f divisible by 3, $A = \mathrm{PSU}_3(2^f)\langle \phi^2 \iota \rangle$, where $\iota \in \mathrm{PGU}_3(q) \setminus \mathrm{PSU}_3(q)$ and ϕ is a field automorphism of order $2f$, $H = \mathrm{PSU}_3(q) \rtimes \langle \phi^6 \rangle$ and $K_M \cap G$ is the stabilizer in G of a decomposition of the 3-dimensional space underlying G into the direct sum of three orthogonal non-singular 1-subspaces.*

Proof The hypothesis that $\{H, K\}$ is a normal 2-covering of A implies that $H \neq A$. In particular, since we are assuming $G \leq H$, we have

$$G \neq A$$

and A/G is a non-identity group.

If A/G is not nilpotent, then the outer automorphism group of G is not nilpotent and hence, using Lemma 12.2, part (1) holds.

---

[2] Here, $\varepsilon = 1$ when $f \equiv 1, 7 \pmod 8$ and $\varepsilon = -1$ when $f \equiv 3, 5 \pmod 8$.

12.2 The Degenerate Normal 2-Coverings of the Second Type

Suppose next that A/G is nilpotent. As $G \le H \le H_M < A$, we have that $H_M \trianglelefteq A$ and A/H_M is cyclic of prime order. Let $\bar{A} := A/G$ and let us adopt the "bar" notation for the projection of A onto $\bar{A}$. As $\{H, K\}$ are components of a normal 2-covering of A, we have

$$\bar{A} = \bigcup_{x \in \bar{A}} \bar{H}^x \cup \bigcup_{x \in \bar{A}} \bar{K}^x. \tag{12.2}$$

We claim that

$$A = KG. \tag{12.3}$$

Note that since H is a proper subgroup of A containing G, we have that $\bar{H}$ is a proper subgroup of $\bar{A}$. If $\bar{A}$ is cyclic, then (12.2) implies $\bar{A} = \bar{H} \cup \bar{K}$ and hence we deduce $\bar{A} = \bar{K}$, that is, (12.3) is satisfied. If $\bar{A}$ is not cyclic, then by Lemma 12.1 $\bar{A}$ admits no normal 2-covering and thus (12.2) implies $\bar{A} = \bar{K}$, that is, (12.3) is satisfied.

If $G \le K_M$, then (12.3) gives $A = KG \le K_M G = K_M$, contradicting the maximality of K_M. Therefore $G \nleq K_M$. Since G is the unique minimal normal subgroup of A, we deduce that K_M is a core-free subgroup of A.

Let Ω be the set of right cosets of K_M in A. As K_M is core-free in A, we may identify A with the primitive permutation group induced by the faithful action of A on Ω. Recall that an element $g \in A$ is said to be a **derangement** if g fixes no point of Ω. Following the notation in [3], we denote by $\mathcal{D}(A)$ the subset of A consisting of the derangements of A and by $D(A)$ the subgroup of A generated by the derangements of A, that is,

$$D(A) := \langle g \in A \mid g \text{ derangement} \rangle.$$

Clearly $D(A) \trianglelefteq A$. Observe that $A \setminus \mathcal{D}(A)$ consists of all the permutations of A fixing some point of Ω and hence

$$A \setminus \mathcal{D}(A) = \bigcup_{g \in A} K_M^g.$$

Therefore,

$$\mathcal{D}(A) = A \setminus \bigcup_{g \in A} K_M^g \subseteq \bigcup_{g \in A} H^g \subseteq H_M. \tag{12.4}$$

As a consequence, $D(A) \le H_M < A$.

Let $\Omega_*^2 := \{(\alpha, \beta) \in \Omega \times \Omega \mid \alpha \ne \beta\}$ and consider the action of A and H_M on Ω_*^2. We claim that no orbit of A on Ω_*^2 coincides with an orbit of H_M on Ω_*^2. Assume, by contradiction, that there exists $(\alpha, \beta) \in \Omega_*^2$ such that $(\alpha, \beta)^{H_M} = (\alpha, \beta)^A$. Let $x \in A \setminus H_M$. Then there exists $y \in H_M$ such that $(\alpha^y, \beta^y) = (\alpha^x, \beta^x)$. Thus we have $\alpha^{xy^{-1}} = \alpha$ and $\beta^{xy^{-1}} = \beta$ so that $xy^{-1} \in A$ fixes at least two points of Ω. Now by the main result [3, Theorem 1.1 (b)] of Bailey, Cameron, Giudici and Royle, all the permutations of A fixing at least two points of Ω belong to $D(A)$. It follows that $xy^{-1} \in D(A) \leq H_M$. Since $y \in H_M$ we deduce the contradiction $x \in H_M$. Recall that A acts primitively on Ω so that its normal subgroup H_M acts transitively on Ω. Hence, within the terminology of Guralnick et al. [54], we have that the pair (A, H_M) is *exceptional* with cyclic quotient A/H_M. Those pairs are classified in [54, Theorem 1.5], which implies[3] that one of the following holds:

(1) G is a group of Lie type and $K_M \cap G$ is the centralizer in G of a field automorphism of odd prime order r. Moreover, r is not the characteristic of G, unless $G = \mathrm{PSL}_2(q)$;
(2) $G = \mathrm{PSL}_2(2^f)$ with $f > 1$ odd and $K_M \cap H_M = D_{2(2^f+1)}$;
(3) $G = \mathrm{PSL}_2(p^f)$, $H_M \in \{G, \mathrm{PGL}_2(p^f)\}$ and $K_M \cap G = D_{p^f-1}$ with p odd and f even;
(4) $G = \mathrm{PSL}_2(3^f)$ and $K_M \cap G = D_{3^f+1}$, with $f \geq 3$ odd;
(5) $G = \mathrm{Sz}(2^f)$ and $K_M \cap G$ is the normalizer of a Sylow 5-subgroup of G;
(6) $G = \mathrm{PSU}_3(2^f)$ with $f > 1$ odd and $K_M \cap G$ is the stabilizer in G of a decomposition of the 3-dimensional space underlying G into the direct sum of three orthogonal non-singular 1-subspaces.

We now deal with each of these six possibilities in turn, starting with the easiest cases. In particular, in part (1), we obtain that part (2a) holds.

Suppose that $G = \mathrm{PSL}_2(2^f)$ and that we are in part (2). Observe that $\mathrm{PSL}_2(2^f) = \mathrm{SL}_2(2^f)$. We aim to prove that part (2b) holds. Now, $\mathrm{Aut}(\mathrm{SL}_2(2^f)) = \mathrm{SL}_2(2^f) \rtimes \langle \phi \rangle$, where ϕ is a field automorphism of odd order $f > 1$. We have $A = \mathrm{SL}_2(2^f) \rtimes \langle \phi^e \rangle$, for some divisor e of f with $e \geq 1$. The nature of K_M is easily found. Indeed, we know that $K_M \not\leq G$ and thus $A = K_M G$, so that $K_M \cong (K_M \cap G).A/G = (K_M \cap G).f/e$. Since $K_M \cap G \leq D_{2(2^f+1)}$, we have that the proper subgroup of A given by $D_{2(2^f+1)}.\langle \phi^e \rangle$ contains K_M and, by the maximality of K_M, we deduce that $K_M \cong D_{2(2^f+1)}.\langle \phi^e \rangle \cong D_{2(2^f+1)}.f/e$. We claim that $A = \mathrm{Aut}(\mathrm{SL}_2(2^f))$. Suppose, by contradiction, that $A < \mathrm{Aut}(\mathrm{SL}_2(2^f))$ so that $e > 1$. As the outer automorphism group of $\mathrm{SL}_2(2^f)$ is cyclic and H_M is a maximal subgroup of A containing $\mathrm{SL}_2(2^f)$, we have $H_M = \mathrm{SL}_2(2^f) \rtimes \langle \phi^{pe} \rangle$, for some prime divisor p of f/e. Let $x \in \mathbf{C}_{\mathrm{SL}_2(2^f)}(\phi^e) = \mathrm{SL}_2(2^e)$ having order $2^e - 1$ and consider $g := x\phi^e \in A$. Now, $g \notin \mathrm{SL}_2(2^f) \rtimes \langle \phi^{pe} \rangle = H_M \trianglelefteq A$ and hence g has a conjugate in K_M. From $x\phi^e \in D_{2(2^f+1)}.\langle \phi^e \rangle$, it follows now that $x \in D_{2(2^f+1)}$, so that $\mathbf{o}(x) = 2^e - 1$ divides $2^f + 1$, which is impossible because e is odd and $e > 1$. Therefore, $A = \mathrm{SL}_2(2^f) \rtimes \langle \phi \rangle = \mathrm{Aut}(\mathrm{SL}_2(2^f))$.

---

[3] Here we are actually slightly weakening the full strength of the theorem of Guralnick, Müller and Saxl. For example, we are including the groups appearing in [54, Theorem 1.5 part (g)] in our part (1).

12.2 The Degenerate Normal 2-Coverings of the Second Type

Now, $H = \mathrm{SL}_2(2^f) \rtimes \langle \phi^e \rangle$, where e is a divisor of f with $e > 1$. Suppose that e is not a prime number and let p be a prime dividing e. Let $x \in \mathbf{C}_{\mathrm{SL}_2(2^f)}(\phi^p) = \mathrm{SL}_2(2^p)$ having order $2^p - 1$ and consider $g := x\phi^p \in A$. Now, $g \notin H = \mathrm{SL}_2(2^f) \rtimes \langle \phi^e \rangle$, because $p < e$. Hence $g = x\phi^p$ has a conjugate in $K_M \cong D_{2(2^f+1)}.\langle \phi \rangle$. It follows that $x \in D_{2(2^f+1)}$, so that $\mathbf{o}(x) = 2^p - 1$ divides $2^f + 1$, which is impossible because $p > 1$ is odd. Therefore, $e = |A : H|$ is prime and thus $H = H_M$ is normal in A.

Suppose that $G = \mathrm{Sz}(2^f)$ and that we are in part (5). Here the argument follows verbatim the proof for the case (2). Set $q := 2^f$. We aim to prove that part (2d) holds. Now, $\mathrm{Aut}(\mathrm{Sz}(q)) = \mathrm{Sz}(q) \rtimes \langle \phi \rangle$, where ϕ is a field automorphism of odd order $f > 1$. We have $A = \mathrm{Sz}(q) \rtimes \langle \phi^e \rangle$, for some divisor e of f with $e > 1$ because $A \neq G$. The nature of K_M is easily found. We know that $K_M \cap G = \mathbf{N}_G(P)$, where P is a Sylow 5-subgroup of G. In particular, $P \neq 1$, because K_M is a proper subgroup of A. Let $\varepsilon = 1$ when $f \equiv 1, 7 \pmod{8}$ and $\varepsilon = -1$ when $f \equiv 3, 5 \pmod{8}$. As

$$q^2 + 1 = (q + \sqrt{2q} + 1)(q - \sqrt{2q} + 1),$$

it is easily seen that 5 divides $q + \varepsilon\sqrt{2q} + 1$ and 5 is relatively prime to $(q - \varepsilon\sqrt{2q} + 1)(q - 1)$. Now, using [9, Table 8.16], we deduce

$$K_M = \mathbf{N}_A(P) \cong (q + \varepsilon\sqrt{2q} + 1) : 4 : \frac{f}{e}.$$

We claim that $A = \mathrm{Aut}(\mathrm{Sz}(q)) = \mathrm{Sz}(q) \rtimes \langle \phi \rangle$. Suppose, by contradiction, that $A < \mathrm{Aut}(\mathrm{Sz}(q))$ so that $e > 1$. As the outer automorphism group of $\mathrm{Sz}(q)$ is cyclic and H_M is a maximal subgroup of A containing $\mathrm{Sz}(q)$, we have $H_M = \mathrm{Sz}(q) \rtimes \langle \phi^{pe} \rangle$, for some prime divisor p of f/e. Let $x \in \mathbf{C}_{\mathrm{Sz}(q)}(\phi^e) = \mathrm{Sz}(2^e)$ having order $2^e - 1$ and consider $g := x\phi^e \in A$. Now, $g \notin \mathrm{Sz}(q) \rtimes \langle \phi^{pe} \rangle = H_M \trianglelefteq A$ and hence g has a conjugate in K_M. From $x\phi^e \in \mathbf{N}_A(P) = ((q + \varepsilon\sqrt{2q} + 1) : 4) : \langle \phi^e \rangle$, it follows now that $x \in (q + \varepsilon\sqrt{2q} + 1) : 4$, so that $\mathbf{o}(x) = 2^e - 1$ divides $q + \varepsilon\sqrt{2q} + 1 = 2^f + \varepsilon 2^{(f+1)/2} + 1$. Since e divides f, $2^e - 1$ divides $2^f - 1$ and hence $2^e - 1$ divides

$$2^f + \varepsilon 2^{\frac{f+1}{2}} + 1 - (2^f - 1) = \varepsilon 2^{\frac{f+1}{2}} + 2.$$

In turn, this implies that $2^e - 1$ divides $\varepsilon 2^{(f-1)/2} + 1$. When $\varepsilon = -1$, from Lemma 2.2 (1), we get

$$2^e - 1 = \gcd(2^e - 1, \varepsilon 2^{(f-1)/2} + 1) = \gcd(2^e - 1, 2^{(f-1)/2} - 1)$$
$$= 2^{\gcd(e, (f-1)/2)} - 1 = 1,$$

because e divides f. Similarly, when $\varepsilon = 1$, from Lemma 2.2 (3), we get

$$2^e - 1 = \gcd(2^e - 1, \varepsilon 2^{(f-1)/2} + 1) = \gcd(2^e - 1, 2^{(f-1)/2} + 1) = \gcd(2, q - 1),$$

because f is odd and q is even. Therefore, $A = \mathrm{Sz}(q) \rtimes \langle \phi \rangle = \mathrm{Aut}(\mathrm{Sz}(q))$.

Now, $H = \mathrm{Sz}(q) \rtimes \langle \phi^e \rangle$, where e is a divisor of f with $e > 1$. If e is not a prime number, then arguing as usual we obtain a contradiction because H and K are not the components of a normal 2-covering. Therefore, $e = |A : H|$ is prime.

Suppose that $G = \mathrm{PSL}_2(3^f)$ and that we are in part (4). Here we prove that no normal 2-covering arises. Let $q := 3^f$. Now, let ι be the projective image of

$$\begin{pmatrix} -1 & 0 \\ 0 & 1 \end{pmatrix} \in \mathrm{GL}_2(q)$$

in $\mathrm{PGL}_2(q)$ and let ϕ be a field automorphism of $\mathrm{PSL}_2(q)$ of order f. Thus $\mathrm{Aut}(G) = \mathrm{Aut}(\mathrm{PSL}_2(q)) = G \rtimes \langle \iota \phi \rangle$, where $\langle \iota \phi \rangle$ is a cyclic group of order $2f$, because f is odd and ι and ϕ commute.

In particular, $A = G \rtimes \langle (\iota\phi)^{e_1} \rangle$, for some divisor e_1 of $2f$, and $H = G \rtimes \langle (\iota\phi)^{e_1 e_2} \rangle$, for some divisor e_2 of $2f/e_1$ with $e_2 > 1$. Assume first that e_1 is even, that is, $e_1 := 2e_1'$, for some divisor e_1' of f. Thus

$$A = G \rtimes \langle (\iota\phi)^{e_1} \rangle = G \rtimes \langle \phi^{e_1'} \rangle, \quad H = G \rtimes \langle \phi^{e_1' e_2} \rangle.$$

Observe that, for every $s \in G$, $\phi^{e_1'} s \in A \setminus H$ and hence $\phi^{e_1'} s$ has a conjugate in K_M. We now apply the Shintani descent with $\sigma := \phi^{e_1'}$, $e := f/e_1'$ and $X := \mathrm{PSL}_2(\bar{\mathbb{F}}_3)$, where $\bar{\mathbb{F}}_3$ is the algebraic closure of $\mathbb{F}_3$. Thus, $\mathcal{G} = \mathrm{PSL}_2(q)$ and $\mathcal{H} = \mathrm{PSL}_2(3^{e_1'})$. Using Lemma 12.4 and the surjectivity of the Shintani descent, we choose $s \in G$ such that $(\sigma s)^e$ has order 3. Since $\sigma s = \phi^{e_1'} s$ has a conjugate in K_M, we have that $(\sigma s)^e \in K_M \cap G = D_{3f+1}$ contradicting the fact that D_{3f+1} has no element of order 3.[4] In particular, in this case no example arises for our main result.

---

[4] Strictly speaking, in this particular case, the Shintani descent is not necessary and we could argue by simply using the norm with respect to a field extension. Consider $q' := 3^{e_1'}$, so that $q = q'^{f/e_1'}$. Let

$$N : \mathbb{F}_q \to \mathbb{F}_{q'}$$

be the norm map of the Galois extension $\mathbb{F}_q/\mathbb{F}_{q'}$. In particular, for every $\alpha \in \mathbb{F}_q$,

$$N(\alpha) = \prod_{\theta \in \mathrm{Gal}(\mathbb{F}_q/\mathbb{F}_{q'})} \alpha^\theta = \prod_{\theta \in \langle \sigma \rangle} \alpha^\theta = \alpha^{\sigma^{f/e_1'-1}} \cdots \alpha^\sigma \alpha.$$

From the surjectivity of the norm, there exists $\alpha \in \mathbb{F}_q$ with $N(\alpha) = 1$. Consider now

$$s := \begin{bmatrix} 1 & \alpha \\ 0 & 1 \end{bmatrix} \in G = \mathrm{PSL}_2(q).$$

12.2 The Degenerate Normal 2-Coverings of the Second Type

Suppose that $G = \mathrm{PSL}_2(p^f)$ and that we are in part (3). Let

$$\iota \in \mathrm{PGL}_2(q) \setminus \mathrm{PSL}_2(q).$$

Let ϕ be a field automorphism of $\mathrm{PSL}_2(q)$ of order f. Thus $\mathrm{Aut}(G) = \mathrm{Aut}(\mathrm{PSL}_2(q)) = G\langle \iota, \phi \rangle$. Note that the outer automorphism group of $\mathrm{PSL}_2(q)$ is isomorphic to $\mathbb{Z}_2 \times \mathbb{Z}_f$, because f is even. Since A is a subgroup of $\mathrm{Aut}(G)$ containing G, we obtain one of the following possibilities

(1) $A = G\langle \iota, \phi^i \rangle$, where i is a divisor of f,
(2) $A = G \rtimes \langle \phi^i \rangle$, where $i \neq f$ is a divisor of f,
(3) $A = G\langle \iota \phi^i \rangle$, where i is a divisor of f with f/i even.

Assume first that we are in case (1) or in case (2). We claim that $\phi^i \in H$. This is obvious when we are dealing with case (1) with $i = f$; therefore, in the proof of this claim, when discussing case (1), we may assume that $i \neq f$. Observe that in case (2), we always have $i \neq f$.

For proving the previous claim, we use the Shintani descent with $e := f/i$ and $\sigma := \phi^i$ establishing a bijection among the conjugacy classes of $G \rtimes \langle \phi^i \rangle \leq A$ contained in $\phi^i G \subseteq A$ and the conjugacy classes in $\mathrm{PSL}_2(p^i)$. Since obviously $\mathrm{PSL}_2(p^i)$ contains an element of order p, there exists $s \in G$ such that $(\phi^i s)^e$ has order p and moreover, by (12.1), we have $(\phi^i s)^e \in G$. Consider now $\phi^i s \in A$. Assume, by contradiction, that $\phi^i s$ belongs to an A-conjugate of K_M. Since $G \trianglelefteq A$ we also have that $(\phi^i s)^e$ belongs to an A-conjugate of $K_M \cap G = D_{q-1}$, contradicting the fact that D_{q-1} contains no element of order p. It follows that $\phi^i s \in H$ and, since $H \geq G$, we deduce that $\phi^i \in H$.

If we are in case (2) this leads to the contradiction $H = A$. Thus case (2) cannot arise. If we are in case (1), this leads to $H \geq G \rtimes \langle \phi^i \rangle$ and, since $G \rtimes \langle \phi^i \rangle$ has index 2 in A, we deduce that $H = H_M = G \rtimes \langle \phi^i \rangle$. We show that this does not give rise to a normal 2-covering of A. We argue by contradiction and we suppose that this is a normal 2-covering. Let $g \in \mathrm{PGL}_2(q) = \langle G, \iota \rangle \leq A$ be a Singer cycle. Clearly, $g \notin H \trianglelefteq A$, because H has no elements having order $q + 1$. Therefore, g has an A-conjugate in K_M. Now, K_M is a maximal subgroup of A with $K_M \cap$

Then $g := \phi^{e'_1} s = \sigma s \in A \setminus H$. In particular, g has a conjugate in K_M. Therefore g^{f/e'_1} has also a conjugate in K_M. Thus $g^{f/e'_1} \in K_M \cap G = D_{q+1}$. Now, by (12.1), we have

$$g^{f/e'_1} = (\sigma s)^{f/e'_1} = s^{\sigma^{f/e'_1 - 1}} \cdots s^{\sigma} s = \begin{bmatrix} 1 & \alpha^{\sigma^{f/e'_1 - 1}} \cdots \alpha^{\sigma} \alpha \\ 0 & 1 \end{bmatrix} = \begin{bmatrix} 1 & 1 \\ 0 & 1 \end{bmatrix}.$$

Hence $K_M \cap G = D_{3^f+1}$ contains an element having order 3, however this is a contradiction.
Assume now that e_1 is odd. Thus

$$A = \mathrm{PSL}_2(q) \rtimes \langle (\iota \phi)^{e_1} \rangle = \mathrm{PSL}_2(q) \rtimes \langle \iota, \phi^{e_1} \rangle = \mathrm{PGL}_2(q) \rtimes \langle \phi^{e_1} \rangle.$$

Now, we may argue exactly as above applying the Shintani descent to $\sigma := \phi^{e_1}$, $e := f/e_1$ and $X := \mathrm{PGL}_2(\mathbb{F}_3)$ and using the fact that $D_{q+1}.2$ does not contain elements of order 3.

$\mathrm{PSL}_2(q) = D_{q-1}$. From [9, Table 8.1], when $q \neq 9$, we have $K_M = \mathbf{N}_A(K_M \cap \mathrm{PSL}_2(q)) = D_{2(q-1)}.(f/i)$. Moreover, as $g \in \mathrm{PGL}_2(q) \trianglelefteq A$, we deduce that g has an A-conjugate in $K_M \cap \mathrm{PGL}_2(q) = D_{2(q-1)}$. However, $D_{2(q-1)}$ has no elements of order $q+1$. The case $q = 9$ also does not give rise to a normal 2-covering; this has been verified with a computer computation.

Finally assume that we are in case (3). Then $A = G\langle \phi^i \iota \rangle$ contains the coset $(\phi^i \iota)^2 G = \phi^{2i} G$. Note that, since i is odd dividing f even we also have that $2i$ divides f. We claim that $\phi^{2i} \in H$.

For that purpose, we use the Shintani descent with $e := f/2i$ and $\sigma := \phi^{2i}$, establishing a bijection among the conjugacy classes of $G \rtimes \langle \phi^{2i} \rangle \leq A$ contained in $\phi^{2i} G \subseteq A$ and the conjugacy classes in $\mathrm{PSL}_2(p^{2i})$. Since obviously $\mathrm{PSL}_2(p^i)$ contains an element of order p, there exists $s \in G$ such that $(\phi^{2i} s)^e$ has order p and moreover, by (12.1), we have $(\phi^{2i} s)^e \in G$. Consider now $\phi^{2i} s \in A$. Assume, by contradiction, that $\phi^{2i} s$ belongs to an A-conjugate in K_M. Since $G \trianglelefteq A$ we also have that $(\phi^{2i} s)^e$ belongs to an A-conjugate of $K_M \cap G = D_{q-1}$, contradicting the fact that D_{q-1} contains no element of order p. It follows that $\phi^{2i} s \in H$ and, since $H \geq G$, we deduce that $\phi^{2i} \in H$.

As a consequence, we have that $H \geq G \langle \phi^{2i} \rangle$ and, since $G \rtimes \langle \phi^{2i} \rangle$ has index 2 in $A = G\langle \iota \phi^i \rangle$, we deduce that $H = H_M = G \rtimes \langle \phi^{2i} \rangle$.

Suppose finally that $G = \mathrm{PSU}_3(2^f)$ and that we are in part (6). We consider the Hermitian form having matrix:

$$\begin{pmatrix} 0 & 0 & 1 \\ 0 & 1 & 0 \\ 1 & 0 & 0 \end{pmatrix}.$$

Set $q := 2^f$ and let λ be a generator of the subgroup of $\mathbb{F}_{q^2}^*$ having order $q+1$. Now, let ι be the projective image of

$$\begin{pmatrix} 1 & 0 & 0 \\ 0 & \lambda & 0 \\ 0 & 0 & 1 \end{pmatrix} \in \mathrm{GU}_3(q)$$

in $\mathrm{PGU}_3(q)$ and let ϕ be a graph-field automorphism of $\mathrm{PSU}_3(q)$ of order $2f$. Thus $\mathrm{Aut}(G) = \mathrm{Aut}(\mathrm{PSU}_3(q)) = G\langle \iota, \phi \rangle$, where the outer automorphism group of $\mathrm{PSU}_3(q)$ is isomorphic to $(\mathbb{Z}_3 \times \mathbb{Z}_f).\mathbb{Z}_2 \cong \mathbb{Z}_3 \times \mathbb{Z}_{2f} \cong S_3 \times \mathbb{Z}_f$, because f is odd. In particular, we have $\iota^\phi = \iota^{-1} \pmod{G}$ and $\iota^3 = 1 \pmod{G}$. Since A is a subgroup of $\mathrm{Aut}(G) = G\langle \iota, \phi \rangle$ containing G, we obtain one of the following possibilities

(1) $A = G\langle \iota, \phi^i \rangle$, for some divisor i of $2f$,
(2) $A = G \rtimes \langle \phi^i \rangle$, for some divisor i of $2f$, with $i \neq 2f$,
(3) $A = G\langle \iota^\varepsilon \phi^i \rangle$, for some $\varepsilon \in \{-1, 1\}$ and for some divisor i of f,
(4) $A = G\langle \iota^\varepsilon \phi^{2i} \rangle$, for some $\varepsilon \in \{-1, 1\}$ and for some divisor i of f with f/i divisible by 3.

12.2 The Degenerate Normal 2-Coverings of the Second Type 171

Observe that in (3), we have

$$(\iota^\varepsilon \phi^i)^2 = \phi^{2i},$$

and that in (4) we have

$$(\iota^\varepsilon \phi^{2i})^3 = \iota^{3\iota} \phi^{6i} \pmod{G} = \phi^{6i} \pmod{G}. \tag{12.5}$$

Moreover, $\iota^{-1}\phi\iota \equiv \phi\iota^2 \pmod{G}$ and hence the groups appearing in (2) with i odd and those in (3) are conjugate under an element in Aut(G).

We claim that

$$\forall j \mid f, \text{ if } \phi^{2j} \in A, \text{ then } \phi^{2j} \in H. \tag{†}$$

Let j be a divisor of f and assume that $\phi^{2j} \in A$. Let $\sigma := \phi^{2j}$ so that $o(\sigma) = f/j =: e$. We have $G \rtimes \langle \sigma \rangle \leq A$ and we apply the Shintani descent to the conjugacy classes of this group contained in the coset $\phi^{2j} G = \sigma G$. Note that the codomain of the descent is given here by $\mathrm{PSU}_3(2^j) \geq \mathrm{PSU}_3(2) \geq Q_8$. Then there exists $s \in G$ such that $(\phi^{2j}s)^e$ has order 4. Consider then $\phi^{2j}s \in A$. If it belongs to K_M up to A-conjugacy, then we have $(\phi^{2j}s)^e$ in $K_M \cap G \cong (2^f + 1)^2.S_3$ up to A-conjugacy, contradicting the fact that the group $(2^f + 1)^2.S_3$ contains no element of order 4.

We next claim that

$$\text{if } \iota \in A, \text{ then } \iota \in H. \tag{††}$$

Let $\iota \in A$. Then $A \supseteq \mathrm{PGU}_3(q)$. Let $g \in \mathrm{PGU}_3(q)$ be a Singer cycle. Assume that $g \in K_M$ up to A-conjugacy. Then

$$g \in K_M \cap \mathrm{PGU}_3(q) \leq (K_M \cap \mathrm{PSU}_3(q)).3 = (q+1)^2.S_3.3$$

up to A-conjugacy. Since $f > 1$, there exists $r \in P_6(q)$, that is r is a primitive prime divisor of $q^6 - 1 = 2^{6f} - 1$, and $r \mid o(g)$. Then $r \nmid q + 1$ and $r \geq 7$ so that it cannot divide the order of $(q+1)^2.S_3.3$ and we get a contradiction. It follows that $\iota \in H$ up to A-conjugacy. It remains to justify that $\iota \in H$. Now, for some $a \in A$, we have $\iota^a \in H$ and hence

$$H \geq \langle G, \iota^a \rangle = \langle G, \iota \rangle^a = \mathrm{PGU}_3(q)^a = \mathrm{PGU}_3(q) = \langle G, \iota \rangle.$$

We consider now the various possibilities for A.

Let A be as in (1). If i is even, then we have $i = 2j$, for some $j \mid f$, and by (†) and (††), we deduce $H = A$, which is a contradiction. If i is odd, then $i \mid f$ and we have $\phi^{2i} \in A$. So, by (†), we deduce $\phi^{2i} \in H$. Moreover, by (††), we have $\iota \in H$. Thus $H \supseteq G\langle \iota, \phi^{2i} \rangle$. Since this last subgroup of A has index 2, we deduce $H = H_M = G\langle \iota, \phi^{2i}\rangle$. Now, let $s \in \mathrm{PGU}_3(q) \le A$ be a Singer cycle and let $g := s\phi^i$. Observe that $g \notin H$ and hence $g \in K_M = (K_M \cap \mathrm{PGU}_3(q)).\langle \phi^i\rangle$ up to $\mathrm{Aut}(A)$-conjugacy. Thus $s \in K_M \cap \mathrm{PGU}_3(q)$. However, arguing as above we reach a contradiction.

Let A be as in (2). If i is even, then we have $i = 2j$ for some $j \mid f$ and, by (†), we get $\phi^i \in H$. Then $H = A$, which is a contradiction. If i is odd, by (†), we get $\phi^{2i} \in H$ so that $H \supseteq G \rtimes \langle \phi^{2i}\rangle$. Since this last subgroup of A has index 2, we deduce $H = H_M = G \rtimes \langle \phi^{2i}\rangle$. Now, let $s \in \mathrm{PSU}_3(q) \le A$ be a Singer cycle and let $g := s\phi^i$. Observe that $g \notin H$ and hence $g \in K_M = (K_M \cap \mathrm{PSU}_3(q)).\langle \phi^i\rangle$ up to $\mathrm{Aut}(A)$-conjugacy. Thus $s \in K_M \cap \mathrm{PSU}_3(q)$ and, arguing as above, we reach a contradiction.

Let A be as in (3). We have shown above that in this case A is $\mathrm{Aut}(G)$-conjugate to a group as in part (2). Therefore, there exists $\tilde{A}$ as in (2) and $\nu \in \mathrm{Aut}(G)$ such that $A^\nu = \tilde{A}$. From $A = \bigcup_{a \in A} H^a \cup \bigcup_{a \in A} K^a$, it follows that

$$\tilde{A} = A^\nu = \bigcup_{a \in A} H^{a\nu} \cup \bigcup_{a \in A} K^{a\nu}.$$

Now, inside $\mathrm{Aut}(G)$, we have $A\nu = \nu\tilde{A}$ and $H^\nu, K^\nu \le \tilde{A}$. Thus we get

$$\tilde{A} = \bigcup_{a \in \tilde{A}} H^{\nu a} \cup \bigcup_{a \in \tilde{A}} K^{\nu a},$$

so that $\tilde{A}$ admits a 2-normal covering. However we have shown that no group of type (2) admits a 2-normal covering.

Let finally A be as in (4). Therefore, $A = G\langle \iota^\varepsilon \phi^{2i}\rangle$, where $\varepsilon \in \{-1, 1\}$ and i is a divisor of f with f/i divisible by 3. By (12.5), we have that $\phi^{6i} \in A$ so that, by (†), $\phi^{6i} \in H$. It follows that $H = H_M = G \rtimes \langle \phi^{6i}\rangle$ has index 3 in A. In order to prove (2e), it suffices to show that $i = 1$.

For each $a \in \mathbb{F}_q$ with $a \ne 0$, the projective image x_a of the matrix

$$\begin{pmatrix} a & 0 & 0 \\ 0 & 1 & 0 \\ 0 & 0 & a^{-1} \end{pmatrix}$$

lies in $\mathrm{PSU}_3(q)$ and hence $\iota^\varepsilon \phi^{2i} x_a \in A$. Clearly, $\iota^\varepsilon \phi^{2i} x_a \notin G \rtimes \langle \phi^{2i}\rangle = H$ and hence $\iota^\varepsilon \phi^{2i} x_a$ has a conjugate in K_M. Now, as ϕ^{2i} commutes with ι and as $\phi^{2f} = 1$, we obtain

$$(\iota^\varepsilon \phi^{2i} x_a)^{f/i} = (\phi^{2i} \iota^\varepsilon x_a)^{f/i} \in \mathrm{PSU}_3(q),$$

12.2 The Degenerate Normal 2-Coverings of the Second Type

which is A-conjugate to an element of $K_M \cap \mathrm{PSU}_3(q) \cong (q+1)^2.S_3$. Set $\sigma := \phi^{2i}$. Now a direct computation using (12.1) shows that the f/i power of $\phi^{2i} \iota^\varepsilon x_a$ is

$$\begin{pmatrix} N(a^\varepsilon) & 0 & 0 \\ 0 & 1 & 0 \\ 0 & 0 & N(a^{-\varepsilon}) \end{pmatrix},$$

where

$$N : \mathbb{F}_q \to \mathbb{F}_{q^{\frac{i}{f}}}$$

is the Galois norm of the field extension $\mathbb{F}_q/\mathbb{F}_{q^{i/f}}$. Since the norm map is surjective, there exists $a \in \mathbb{F}_q$ such that $N(a)$ has order $q^{i/f} - 1 = 2^i - 1$. This proves that $K_M \cap \mathrm{PSU}_3(q) \cong (q+1)^2.S_3$ contains an element having order $2^i - 1$. However, recalling that i is odd, we have that $2^i - 1$ is relatively prime to $(q+1)^2.|S_3|$. Thus $i = 1$, which is what we wanted. □

All good things in life at some point come to an end. Therefore, also in our work, we need to wrap up and finish. We close this section observing that Theorems 12.3 and 12.5 are far from giving a complete classification of the degenerate normal 2-coverings of almost simple groups. However, we believe that they could be the beginning of interesting new research.

Actually, we cannot resist from making one last observation concerning Theorem 12.5. Indeed, in Theorem 12.5 part (2a), for a suitable prime power q, every group of Lie type can arise. Rather than giving full details we just give one example using projective special linear groups.

Let n be a positive integer with $n \geq 2$ and let p be a prime number. Now, let f be a prime number with f relatively prime to $p^i - 1$, for every $i \in \{1, \ldots, n\}$. For instance, we may take f to be any prime number greater than $p^n - 1$. Next, let $q := p^f$, $G := \mathrm{PSL}_n(q)$, let $\phi \in \mathrm{Aut}(G)$ be a field automorphism of order f and let $A := G \rtimes \langle \phi \rangle$.

Observe that f is relatively prime to $|G|$: indeed, from Fermat's little theorem, for every $i \in \{1, \ldots, n\}$, we have $q^i = p^{if} \equiv p^i \pmod{f}$. Since f is relatively prime to $p^i - 1$, we deduce that also f is relatively prime to $q^i - 1$.

Now, let $H := G = \mathrm{PSL}_n(q)$ and let $K := \mathbf{C}_A(\phi) = \mathrm{PSL}_n(p) \times \langle \phi \rangle$. We claim that $\{H, K\}$ is a normal 2-covering of A. Let $g \in A$. If $g \in G$, then $g \in H$, because $H = G$. If $g \notin G$, then $g = s\phi^i$, where $i \in \{1, \ldots, f-1\}$ and $s \in G$. Since $g \pmod G$ has order f in A/G, $\mathbf{o}(g)$ is divisible by f. Set $x := g^{\mathbf{o}(g)/f}$ and observe that x has order f. By Sylow's theorem, $\langle x \rangle$ and $\langle \phi \rangle$ are A-conjugate, and hence there exists $a \in A$ with $\langle x \rangle^a = \langle \phi \rangle$. Now,

$$g^a \in (\mathbf{C}_A(g^{\mathbf{o}(g)/f}))^a = (\mathbf{C}_A(x))^a = \mathbf{C}_A(x^a) = \mathbf{C}_A(\phi) = K.$$

References

1. Aschbacher, M.: On the maximal subgroups of the finite classical groups. Invent. Math. **76**, 469–514 (1984)
2. Baer, R.: Supersoluble groups. Proc. Am. Math. Soc. **6**, 16–32 (1955)
3. Bailey, R.A., Cameron, P.J., Giudici, M., Royle, G.F.: Groups generated by derangements. J. Algebra **572**, 245–262 (2021)
4. Barraclough, R.W., Wilson, R.A.: The character table of a maximal subgroup of the Monster. LMS J. Comput. Math. **10**, 161–175 (2007)
5. Bereczky, A.: Maximal overgroups of Singer elements in classical groups. J. Algebra **234**, 187–206 (2000)
6. Bosma, W., Cannon, J., Playoust, C.: The Magma algebra system. I. The user language. J. Symbolic Comput. **24**(3–4), 235–265 (1997)
7. Boston, N., Dabrowski, W., Foguel, T., Gies, P.J., Leavitt, J., Ose, D.T.: The proportion of fixed-point-free elements in a transitive group. Commun. Algebra **21**, 3259–3275 (1993)
8. Brandl, R., Bubboloni, D., Hupp, I.: Polynomials with roots $\bmod p$ for all primes p. J. Group Theory **4**, 233–239 (2001)
9. Bray, J.N., Holt, D.F., Roney-Dougal, C.M.: The Maximal Subgroups of the Low Dimensional Classical Groups. London Mathematical Society Lecture Note Series, vol. 407. Cambridge University Press, Cambridge (2013)
10. Breuer, T.: Manual for the GAP Character Table Library, Version 1.1. RWTH Aachen (2004)
11. Breuer, T.: Four Primitive Permutation Characters of the Monster Group. https://www.math.rwth-aachen.de/~Thomas.Breuer/ctbllib/doc2/chap8.html#X8337F3C682B6BE63
12. Britnell, J.R., Maróti, A.: Normal coverings of linear groups. Algebra Number Theory **7**(9), 2085–2102 (2013)
13. Bubboloni, D.: Coverings of the symmetric and alternating groups. Dipartimento di Matematica "U. Dini" - Università di Firenze 7 (1998)
14. Bubboloni, D., Lucido, M.S.: Coverings of linear groups. Commun. Algebra **30**, 2143–2159 (2002)
15. Bubboloni, D., Sonn, J.L.: Intersective S_n polynomials with few irreducible factors. Manuscr. Math. **151**, 477–492 (2016)
16. Bubboloni, D., Lucido, M.S., Weigel, Th.: Generic 2-coverings of finite groups of Lie-type. Rend. Sem. Mat. Padova **115**, 209–252 (2006)
17. Bubboloni, D., Praeger, C.E., Spiga, P.: Normal coverings and pairwise generation of finite alternating and symmetric groups. J. Algebra **390**, 199–215 (2013)

18. Bubboloni, D., Praeger, C.E., Spiga, P.: Linear bounds for the normal covering number of the symmetric and alternating groups. Monatsh. Math. **191**, 229–247 (2020)
19. Burness, T.C., Giudici, M.: Classical Groups, Derangements and Primes. Australian Mathematical Society Lecture Series, vol. 25. Cambridge University Press, Cambridge (2016)
20. Burness, T.C., Guest, S.: On the uniform spread of almost simple linear groups. Nagoya Math. J. **209**, 35–109 (2013)
21. Burness, T.C., Tong-Viet, H.P.: Primitive permutation groups and derangements of prime power order. Manuscr. Math. **150**, 255–291 (2016)
22. Burness, T.C., O'Brien, E.A., Wilson, R.A.: Base sizes for sporadic simple groups. Israel J. Math. **177**, 307–333 (2010)
23. Buturlakin, A.A.: Spectra of finite linear and unitary groups. Algebra Logica **47**, 157–173 (2008)
24. Buturlakin, A.A.: Spectra of finite symplectic and orthogonal groups. Mat. Tr. **13**, 33–83 (2010)
25. Buturlakin, A.A., Grechkoseeva, M.A.: The cyclic structure of maximal tori in finite classical groups. Algebra Logic **46**, 73–89 (2007)
26. Cameron, P.J.: Projective and Polar Spaces. QMW Maths Notes, vol. 13. Published by the School of Mathematical Sciences. Queen Mary and Westfield College, Mile End Road, London (1992)
27. Carter, R.W.: Finite Groups of Lie Type: Conjugacy Classes and Complex Characters. A Wiley-Interscience Publication. Wiley, New York (1985)
28. Cline, E., Parshall, B., Scott, L.: Cohomology of finite groups of Lie type, I. Inst. Hautes Ètudes Sci. Publ. Math. **45**, 169–191 (1975)
29. Conway, J.H., Curtis, R.T., Norton, S.P., Parker, R.A., Wilson, R.A.: An ATLAS of Finite Groups. Clarendon Press, Oxford (1985). Reprinted with corrections 2003
30. De Franceschi, G.: Centralizers and conjugacy classes in finite classical groups. PhD thesis, University of Auckland (2018)
31. De Franceschi, G.: Centralizers and conjugacy classes in finite classical groups. arXiv:2008.12651 [math.GR]
32. Dietrich, H., Lee, M., Popiel, T.: The maximal subgroups of the Monster. arXiv:2304.14646 [math.GR]
33. Dye, R.H.: Interrelations of symplectic and orthogonal groups in characteristic two. J. Algebra **59**, 202–221 (1979)
34. Enomoto, H.: The conjugacy classes of Chevalley groups of type (G2) over finite fields of characteristic 2 or 3. J. Fac. Sci. Univ. Tokyo Sect. I **16**, 497–512 (1969)
35. Enomoto, H.: The characters of the finite symplectic group $Sp(4, q)$, $q = 2^f$. Osaka Math. J. **9**, 75–94 (1972)
36. Enomoto, H., Yamada, H.: The characters of $G_2(2^n)$. Jpn. J. Mat **12**, 325–377 (1986)
37. Fulman, J., Guralnick, R.M.: Derangements in simple and primitive groups. In: Groups, Combinatorics, and Geometry (Durham, 2001), pp. 99–121. World Sci. Publ., River Edge (2003)
38. Fulman, J., Guralnick, R.M.: Bounds on the number and sizes of conjugacy classes in finite Chevalley groups with applications to derangements. Trans. Am. Math. Soc. **364**, 3023–3070 (2012)
39. Fulman, J., Guralnick, R.M.: Derangements in subspace actions of finite classical groups. Trans. Am. Math. Soc. **369**(4), 2521–2572 (2017)
40. Fulman, J., Guralnick, R.M.: Derangements in finite classical groups for actions related to extension field and imprimitive subgroups and the solution of the Boston–Shalev conjecture. Trans. Am. Math. Soc. **370**, 4601–4622 (2018)
41. Fusari, M., Previtali, A., Spiga, P.: Groups having minimal covering number 2 of diagonal type. Math. Nach. 1–9 (2024). https://doi.org/10.1002/mana.202400096
42. Garonzi, M., Lucchini, A.: Covers and normal covers of finite groups. J. Algebra **422**, 148–165 (2015)
43. Garzoni, D.: The invariably generating graph of the alternating and symmetric groups. J. Group Theory **23**, 1081–1102 (2020)

44. Garzoni, D.: Connected components in the invariably generating graph of a finite group. Bull. Aust. Math. Soc. **104**, 453–463 (2021)
45. Garzoni, D., Lucchini, A.: Minimal invariable generating sets. J. Pure Appl. Algebra **224**, 218–238 (2020)
46. Garzoni, D., McKemmie, E.: On the probability of generating invariably a finite simple group. arXiv:2008.03812v2 [math.GR]
47. Gill, N., Giudici, M., Spiga, P.: A generalization of Szep's conjecture for almost simple groups. Vietnam J. Math. (2023). s10013-023-00635-1
48. Godsil, C., Meagher, K.: Erdős–Ko–Rado Theorems: Algebraic Approaches. Cambridge Studies in Advanced Mathematics, vol. 149. Cambridge University Press, Cambridge (2016)
49. Gorenstein, D., Lyons, R., Solomon, R.: The Classification of the Finite Simple Groups, Number 3. Amer. Math. Soc. Surveys and Monographs, vol. 40, 3. American Mathematical Society, Providence (1998)
50. Guest, S., Previtali, A., Spiga, P.: A remark on the permutation representations afforded by the embeddings of $O_{2m}^{\pm}(2^f)$ in $Sp_{2m}(2^f)$. Bull. Aust. Math. Soc. **89**, 331–336 (2014)
51. Guest, S., Morris, J., Praeger, C.E., Spiga, P.: On the maximum orders of elements of finite almost simple groups and primitive permutation groups. Trans. Am. Math. Soc. **367**(11), 7665–7694 (2015)
52. Guralnick, R.M., Malle, G.: Simple groups admit Beauville structures. J. Lond. Math. Soc. **85**(3), 694–721 (2012)
53. Guralnick, R.M., Penttila, T., Praeger, C.E., Saxl, J.: Linear groups with orders having certain large prime divisors. Proc. Lond. Math. Soc. (3) **78**, 167–214 (1999)
54. Guralnick, R.M., Müller, P., Saxl, J.: The rational function analogue of a question of Schur and exceptionality of permutation representations. Mem. Am. Math. Soc. **162**, 1–79 (2003)
55. Hardy, G.H., Wright, E.M.: An Introduction to the Theory of Numbers. Oxford University Press, Oxford (1938)
56. Harper, S., Shintani descent, simple groups and spread. J. Algebra **578**, 319–355 (2021)
57. Hestenes, M.D.: Singer groups. Can. J. Math. **XXII**(3), 492–513 (1970)
58. Higman, G., Neumann, B.H., Neumann, H.: Embedding theorems for groups. J. Lond. Math. Soc. (1) **24**, 247–254 (1949)
59. Huppert, B.: Endliche Gruppen I, Grundlehren der mathematischen Wissenschaften, vol. 134. Springer, Berlin (1967)
60. Huppert, B.: Singer-Zyklen in klassischen Gruppen. Math. Z. **117**, 141–150 (1970)
61. Jehne, W.: Kronecker classes of algebraic number fields. J. Number Theory **9**, 279–320 (1977)
62. Kantor, W.M., Seress, Á.: Large element orders and the characteristic of Lie-type simple groups. J. Algebra **322**, 802–832 (2009)
63. Kantor, W.M., Lubotzky, A., Shalev, A.: Invariable generation and the Chebotarev invariant of a finite group. J. Algebra **348**, 302–314 (2011)
64. Kawanaka, N.: On the irreducible characters of the finite unitary groups. J. Math. Soc. Jpn. **29**, 425–450 (1977)
65. Khukhro, E.I., Mazurov, V.D.: Unsolved Problems in Group Theory. The Kourovka Notebook, No. 20 (2023). arXiv:1401.0300v24
66. Kleidman, P.B., Liebeck, M.W.: The Subgroup Structure of the Finite Classical Groups. London Math. Soc. Lecture Notes, vol. 129. Cambridge University Press, Cambridge (1990)
67. Klingen, N.: Zahlkörper mit gleicher Primzerlegung. J. Reine Angew. Math. **299**, 342–384 (1978)
68. Klingen, N.: Arithmetical Similarities, Prime decompositions and Finite Group Theory. Oxford Mathematical Monographs. Clarendon Press, Oxford (1998)
69. Landau, E.: Über die Maximalordnung der Permutationen gegebenen Grades. Arch. Math. Phys. **5**, 92–103 (1903)
70. Li, C.H., Song, S.J., Raghu Tej Pantangi, V.: Erdős-Ko-Rado problems for permutation groups (2020). arXiv preprint arXiv:2006.10339

71. Liebeck, M.W., Praeger, C.E., Saxl, J.: The Maximal Factorizations of the Finite Simple Groups and Their Automorphism Groups. Memoirs Amer. Math. Soc., vol. 86, number 432. American Mathematical Society, Providence (1990)
72. Liebeck, M.W., Praeger, C.E., Saxl, J.: On factorizations of almost simple groups. J. Algebra **185**(2), 409–419 (1996)
73. Lucido, M.S.: On the n-covers of exceptional groups of Lie type. In: Groups St. Andrews 2005, vol. 2, pp. 621–623. London Math. Soc. Lecture Note Ser., vol. 340. Cambridge University Press, Cambridge (2007)
74. Malle, G., Saxl, J., Weigel, Th.: Generation of classical groups. Geom. Dedic. **49**, 85–116 (1994)
75. Massias, J.P., Nicolas, J.L., Robin, G.: Effective bounds for the maximal order of an element in the symmetric Group. Math. Comput. **53**, 665–678 (1989)
76. Meagher, K., Sarobidy Razafimahatratra, A., Spiga, P.: On triangles in derangement graphs. J. Combin. Theory Ser. A **180**, 26 pp. (2021). Paper No. 105390
77. Pellegrini, M.A.: 2-Coverings for exceptional and sporadic simple groups. Arch. Math. (Basel) **101**, 201–206 (2013)
78. Perlis, R.: On the equation $\zeta_K(s) = \zeta_{K'}(s)$. J. Number Theory **9**, 342–360 (1977)
79. Praeger, C.E.: Covering subgroups of groups and Kronecker classes of fields. J. Algebra **118**, 455–463 (1988)
80. Praeger, C.E.: On octic extensions and a problem in group theory. In: Cheng, K.N., Leong, Y.K. (eds.) Proceedings of the 1987 Singapore Group Theory Conference, pp. 443–463. De Gruyter, Berlin (1989)
81. Praeger, C.E.: Kronecker classes of field extensions of small degree. J. Austral. Math. Soc. Ser. A **50**, 297–315 (1991)
82. Praeger, C.E.: Kronecker classes of fields and covering subgroups of finite groups. J. Austral. Math. Soc. Ser. A **57**, 17–34 (1994)
83. Rabayev, D., Sonn, J.: On Galois realizations of the 2-coverable symmetric and alternating groups. Commun. Algebra **42**, 253–258 (2014)
84. Robinson, D.J.S.: A Course in the Theory of Groups, 2nd edn. Graduate Texts in Mathematics, vol. 80. Springer, Berlin (1996)
85. Saxl, J.: On a question of W. Jehne concerning covering subgroups of groups and Kronecker classes of fields, J. Lond. Math. Soc. (2) **38**, 243–249 (1988)
86. Serre, J.-P.: Galois Cohomology. Springer Monographs in Mathematics. Springer, Berlin (1997)
87. Shalev, A.: A theorem on random matrices and some applications. J. Algebra **199**, 124–141 (1998)
88. Spiga, P.: The Erdős–Ko–Rado theorem for the derangement graph of the projective general linear group acting on the projective space. J. Combin. Theory Ser. A **166**, 59–90 (2019)
89. The GAP Group, GAP – Groups, Algorithms, and Programming, Version 4.11.1 (2021). https://www.gap-system.org
90. Wall, G.E.: On the conjugacy classes in the unitary, symplectic and orthogonal groups. J. Austral. Math. Soc. **3**, 1–62 (1963)
91. Wall, G.E.: Conjugacy classes in projective and special linear groups. Bull. Austral. Math. Soc. **22**, 339–364 (1980)
92. Wilson, R.A.: Maximal subgroups of sporadic groups. In: Finite Simple Groups: Thirty Years of the Atlas and Beyond. Contemp. Math., vol. 694, pp. 57–72. American Mathematical Society, Providence (2017)
93. Wilson, R.A.: The uniqueness of $PSU_3(8)$ in the Monster. Bull. Lond. Math. Soc. **49**, 877–880 (2017)
94. Zsigmondy, K.: Zur Theorie der Potenzreste. Monathsh. für Math. u. Phys. **3**, 265–284 (1892)

LECTURE NOTES IN MATHEMATICS   Springer

Editors in Chief: J.-M. Morel, B. Teissier;

Editorial Policy

1. Lecture Notes aim to report new developments in all areas of mathematics and their applications – quickly, informally and at a high level. Mathematical texts analysing new developments in modelling and numerical simulation are welcome.

   Manuscripts should be reasonably self-contained and rounded off. Thus they may, and often will, present not only results of the author but also related work by other people. They may be based on specialised lecture courses. Furthermore, the manuscripts should provide sufficient motivation, examples and applications. This clearly distinguishes Lecture Notes from journal articles or technical reports which normally are very concise. Articles intended for a journal but too long to be accepted by most journals, usually do not have this "lecture notes" character. For similar reasons it is unusual for doctoral theses to be accepted for the Lecture Notes series, though habilitation theses may be appropriate.

2. Besides monographs, multi-author manuscripts resulting from SUMMER SCHOOLS or similar INTENSIVE COURSES are welcome, provided their objective was held to present an active mathematical topic to an audience at the beginning or intermediate graduate level (a list of participants should be provided).

   The resulting manuscript should not be just a collection of course notes, but should require advance planning and coordination among the main lecturers. The subject matter should dictate the structure of the book. This structure should be motivated and explained in a scientific introduction, and the notation, references, index and formulation of results should be, if possible, unified by the editors. Each contribution should have an abstract and an introduction referring to the other contributions. In other words, more preparatory work must go into a multi-authored volume than simply assembling a disparate collection of papers, communicated at the event.

3. Manuscripts should be submitted either online at www.editorialmanager.com/lnm to Springer's mathematics editorial in Heidelberg, or electronically to one of the series editors. Authors should be aware that incomplete or insufficiently close-to-final manuscripts almost always result in longer refereeing times and nevertheless unclear referees' recommendations, making further refereeing of a final draft necessary. The strict minimum amount of material that will be considered should include a detailed outline describing the planned contents of each chapter, a bibliography and several sample chapters. Parallel submission of a manuscript to another publisher while under consideration for LNM is not acceptable and can lead to rejection.

4. In general, **monographs** will be sent out to at least 2 external referees for evaluation.

   A final decision to publish can be made only on the basis of the complete manuscript, however a refereeing process leading to a preliminary decision can be based on a pre-final or incomplete manuscript.

   Volume Editors of **multi-author works** are expected to arrange for the refereeing, to the usual scientific standards, of the individual contributions. If the resulting reports can be forwarded to the LNM Editorial Board, this is very helpful. If no reports are forwarded or if other questions remain unclear in respect of homogeneity etc, the series editors may wish to consult external referees for an overall evaluation of the volume.

5. Manuscripts should in general be submitted in English. Final manuscripts should contain at least 100 pages of mathematical text and should always include

   – a table of contents;
   – an informative introduction, with adequate motivation and perhaps some historical remarks: it should be accessible to a reader not intimately familiar with the topic treated;
   – a subject index: as a rule this is genuinely helpful for the reader.
   – For evaluation purposes, manuscripts should be submitted as pdf files.

6. Careful preparation of the manuscripts will help keep production time short besides ensuring satisfactory appearance of the finished book in print and online. After acceptance of the manuscript authors will be asked to prepare the final LaTeX source files (see LaTeX templates online: https://www.springer.com/gb/authors-editors/book-authors-editors/manuscriptpreparation/5636) plus the corresponding pdf- or zipped ps-file. The LaTeX source files are essential for producing the full-text online version of the book, see http://link.springer.com/bookseries/304 for the existing online volumes of LNM). The technical production of a Lecture Notes volume takes approximately 12 weeks. Additional instructions, if necessary, are available on request from lnm@springer.com.

7. Authors receive a total of 30 free copies of their volume and free access to their book on SpringerLink, but no royalties. They are entitled to a discount of 33.3 % on the price of Springer books purchased for their personal use, if ordering directly from Springer.

8. Commitment to publish is made by a *Publishing Agreement*; contributing authors of multiauthor books are requested to sign a *Consent to Publish form*. Springer-Verlag registers the copyright for each volume. Authors are free to reuse material contained in their LNM volumes in later publications: a brief written (or e-mail) request for formal permission is sufficient.

Addresses:
Professor Jean-Michel Morel, CMLA, École Normale Supérieure de Cachan, France
E-mail: moreljeanmichel@gmail.com

Professor Bernard Teissier, Equipe Géométrie et Dynamique,
Institut de Mathématiques de Jussieu – Paris Rive Gauche, Paris, France
E-mail: bernard.teissier@imj-prg.fr

Springer: Ute McCrory, Mathematics, Heidelberg, Germany,
E-mail: lnm@springer.com

SPRINGER NATURE

GPSR Compliance

The European Union's (EU) General Product Safety Regulation (GPSR) is a set of rules that requires consumer products to be safe and our obligations to ensure this.

If you have any concerns about our products, you can contact us on ProductSafety@springernature.com

In case Publisher is established outside the EU, the EU authorized representative is:

Springer Nature Customer Service Center GmbH
Europaplatz 3
69115 Heidelberg, Germany

The manufacturer's authorised representative in the EU is Springer Nature Customer Service Centre GmbH, Europaplatz 3, 69115 Heidelberg, Germany. If you have any concerns regarding our products, please contact ProductSafety@springernature.com

Printed and bound by CPI Group (UK) Ltd, Croydon, CR0 4YY

25/03/2026

02078187-0005